Meat Culture

Human-Animal Studies

Series Editor

Kenneth Shapiro (*Animals & Society Institute, USA*)

Editorial Board

Ralph Acampora (*Hofstra University, USA*)
Hilda Kean (*Ruskin College, Oxford, UK*)
Randy Malamud (*Georgia State University, USA*)
Gail Melson (*Purdue University, USA*)
Leslie Irvine (*University of Colorado, USA*)

VOLUME 17

The titles published in this series are listed at *brill.com/has*

Meat Culture

Edited by

Annie Potts

BRILL

LEIDEN | BOSTON

Cover Illustration: Nicolas Lampert, "Attention Chicken!" (Public intervention at the Big Lots in Milwaukee, Wisconsin 2006.) Sculpture by Nicolas Lampert with assistance by Micaela O'Herlihy. http://www.nicolas lampert.org/.

Library of Congress Cataloging-in-Publication Data

Names: Potts, Annie, 1965– editor.
Title: Meat culture / edited by Annie Potts.
Description: Leiden ; Boston : Brill, 2016. | Series: Human-animal studies,
   ISSN 1573-4226 ; Volume 17 | Includes index.
Identifiers: LCCN 2016024323 (print) | LCCN 2016037652 (ebook) | ISBN
   9789004325845 (hardback : alk. paper) | ISBN 9789004325869 (pbk. : alk.
   paper) | ISBN 9789004325852 (E-book)
Subjects: LCSH: Meat—Social aspects. | Meat—Moral and ethical aspects. |
   Meat industry and trade—Social aspects. | Meat industry and trade—Moral
   and ethical aspects.
Classification: LCC GT2868.5 .M35 2016 (print) | LCC GT2868.5 (ebook) | DDC
   641.3/6—dc23
LC record available at https://lccn.loc.gov/2016024323

Typeface for the Latin, Greek, and Cyrillic scripts: "Brill". See and download: brill.com/brill-typeface.

ISSN 1573-4226
ISBN 978-90-04-32586-9 (paperback)
ISBN 978-90-04-32585-2 (e-book)

This paperback is also published in hardback under ISBN 978-90-04-32584-5

Copyright 2017 by Koninklijke Brill NV, Leiden, The Netherlands.
Koninklijke Brill NV incorporates the imprints Brill, Brill Hes & De Graaf, Brill Nijhoff, Brill Rodopi and Hotei Publishing.
All rights reserved. No part of this publication may be reproduced, translated, stored in a retrieval system, or transmitted in any form or by any means, electronic, mechanical, photocopying, recording or otherwise, without prior written permission from the publisher.
Authorization to photocopy items for internal or personal use is granted by Koninklijke Brill NV provided that the appropriate fees are paid directly to The Copyright Clearance Center, 222 Rosewood Drive, Suite 910, Danvers, MA 01923, USA. Fees are subject to change.

This book is printed on acid-free paper and produced in a sustainable manner.

Contents

Acknowledgements VII
List of Contributors VIII

1 What is Meat Culture? 1
    Annie Potts

2 Derrida and *The Sexual Politics of Meat* 31
    Carol J. Adams and Matthew Calarco

3 Rotten to the Bone: Discourses of Contamination and Purity in the European Horsemeat Scandal 54
    Nik Taylor and Jordan McKenzie

4 Live Exports, Animal Advocacy, Race and 'Animal Nationalism' 73
    Jacqueline Dalziell and Dinesh Joseph Wadiwel

5 *The Whopper Virgins*: Hamburgers, Gender, and Xenophobia in Burger King's Hamburger Advertising 90
    Vasile Stănescu

6 With Care for Cows and a Love for Milk: Affect and Performance in Swedish Dairy Industry Marketing Strategies 109
    Tobias Linné and Helena Pedersen

7 "Peace and Quiet and Open Air": *The Old Cow Project* 129
    Melissa Boyde

8 "Do You Know Where the Light Is?" Factory Farming and Industrial Slaughter in Michel Faber's *Under the Skin* 149
    Kirsty Dunn

9 Down on the Farm: Why do Artists Avoid 'Farm' Animals as Subject Matter? 163
    Yvette Watt

10 The Provocative Elitism of 'Personhood' for Nonhuman Creatures in Animal Advocacy Parlance and Polemics   184
    Karen Davis

11 "I Need Fish Fingers and Custard": The Irruption and Suppression of Vegan Ethics in *Doctor Who*   198
    Matthew Cole and Kate Stewart

12 On Ambivalence and Resistance: Carnism and Diet in Multi-species Households   222
    Erika Cudworth

13 Negotiating Social Relationships in the Transition to Vegan Eating Practices   243
    Richard Twine

14 Critical Ecofeminism: Interrogating 'Meat,' 'Species,' and 'Plant'   264
    Greta Gaard

    Index   289

Acknowledgements

Annie Potts sincerely thanks the authors, each of whom inspires her professionally and personally. It has been a pleasure and an honour to work alongside such compassionate and effective animal advocates. She is very grateful to Ken Shapiro, Randy Malamud, Meghan Connolly, Jennifer Pavelko and Maria Baluch for their excellent advice and warm professionalism throughout the editing and publishing process; Philip Armstrong for his helpful review of chapter one; two anonymous readers for their insightful feedback on the full manuscript; and Donelle Gadenne for her meticulous copyediting of all chapters.

This volume respectfully remembers the multispecies sufferers and victims of meat culture.

The following individuals and organizations are kindly acknowledged for their permissions to reproduce diagrams or images:

Chapter 1: 'Numbers of Animals Slaughtered 2011', from *Meat Atlas*, courtesy of Heinrich Böll Foundation and Friends of the Earth Europe; 'New Zealand Dairy Contaminated with Cruelty", courtesy of Save Animals From Exploitation (SAFE), New Zealand; 'Rescued broiler chicks', courtesy of Kay Evans, Chocowinity Chicken Sanctuary, North Carolina, USA.

Chapter 7: 'Lost and Found', courtesy of Munemasa Takahashi, Director *Lost and Found*; 'Paleo-camera', courtesy of Matt Gatton; 'Meat . . . with care!' courtesy of State Records New South Wales, Australia; 'Homebush Abbatoir, 1931', 'Homebush Abbatoir, 1937', 'Beef carcasses, 1954', courtesy of Collection of the State Library of New South Wales, Australia; image from *Old Cow Project* series, courtesy of artist Derek Kreckler.

All other images courtesy of authors.

List of Contributors

Carol J. Adams
is the author of *The Sexual Politics of Meat*, released in 2015 in a 25th anniversary edition as part of Bloomsbury's Revelations series. She is also the author of many other books, most recently *Never Too Late to Go Vegan: The Over-50 Guide to Adopting and Thriving on a Plant-Based Diet* and the co-edited volume (with Lori Gruen), *Ecofeminism: Feminist Intersections with Other Animals and the Earth*. She has been involved in social justice activism including on behalf of the other animals for more than forty years. www.caroljadams.com.

Melissa Boyde
is a Senior Research Fellow in the School of the Arts, English and Media at the University of Wollongong. She is the editor of *Animal Studies Journal* and co-editor of the *Animal Publics* book series, Sydney University Press. As well as her work in animal studies, Melissa is a curator and researcher in modernist art and literature. She is the editor of *Captured: the Animal Within Culture* (Palgrave McMillian, 2014).

Matthew Calarco
is Associate Professor of Philosophy at California State University, Fullerton. He works in the fields of animal philosophy, environmental philosophy, and Continental philosophy. He is the author of *Zoographies: The Question of the Animal from Heidegger to Derrida* (Columbia University Press, 2008). His most recent book is called *Thinking through Animals: Identity, Difference, Indistinction*.

Matthew Cole
is an Honorary Associate and Associate Lecturer with The Open University in the UK. His research centres on Critical Animal Studies and the sociology of human-nonhuman animal relations, including the childhood socialization of human domination, the cultural representation of vegans and veganism and the genealogy of modern veganism. In 2014 he published his first book with Dr Kate Stewart: *Our Children and Other Animals: The Cultural Construction of Human-Animal Relations in Childhood*.

Erika Cudworth
is Professor of Feminist Animal Studies at the University of East London, UK. Her research interests include complexity theory, gender, and human relations with non-human animals, particularly theoretical and political

challenges to exclusive humanism. She is author of *Environment and Society* (Routledge, 2003), *Developing Ecofeminist Theory: The Complexity of Difference* (Palgrave, 2005) and *Social Lives with Other Animals: Tales of Sex, Death and Love* (Palgrave, 2011); co-author of *The Modern State: Theories and Ideologies* (Edinburgh University Press, 2007) and *Posthuman International Relations: Complexity, Ecologism and International Politics* (Zed, 2011); and co-editor of *Technology, Society and Inequality: New Horizons and Contested Futures* (Peter Lang, 2013) and *Anarchism and Animal Liberation: Essays on Complementary Elements of Total Liberation* (McFarland, 2015). Erika's current projects are on animal companions, animals and war, and posthuman emancipation.

Jacqueline Dalziell
is completing a PhD in Sociology at The University of New South Wales in Sydney, Australia. Her work can be broadly construed as a feminist, posthumanist intervention. She is currently focusing her research on the surrealist thinker Roger Caillois and his scholarship on mimesis.

Karen Davis, PhD
is the President and Founder of United Poultry Concerns, a nonprofit organization that promotes the compassionate and respectful treatment of domestic fowl. Her books include *Prisoned Chickens, Poisoned Eggs: An Inside Look at the Modern Poultry Industry*; *More Than a Meal: The Turkey in History, Myth, Ritual, and Reality*; and *The Holocaust and the Henmaid's Tale: A Case for Comparing Atrocities*. Award-winningly profiled in *The Washington Post*, and the author of many published articles and book chapters, Karen maintains a sanctuary for chickens in Virginia, and was inducted into the U.S. Animal Rights Hall of Fame in 2002 for "outstanding contributions to animal liberation." http://www.upc-online.org/karenbio.htm.

Greta Gaard
is Professor of English and Coordinator of the Sustainability Faculty Fellows at University of Wisconsin-River Falls. Her work emerges from the intersections of feminism, environmental justice, queer studies and critical animal studies, exploring a wide range of issues, from interspecies justice, material perspectives on fireworks and space exploration, postcolonial ecofeminism, and the eco-politics of climate change. Author or editor of five books, Gaard's most recent volume is *International Perspectives in Feminist Ecocriticism* (Routledge, 2013), co-edited with Simon Estok and Serpil Oppermann. Her creative nonfiction eco-memoir, *The Nature of Home* (Arizona University Press, 2007), is being translated into Chinese and Portuguese.

Tobias Linné
holds a Ph.D. in Sociology and is Assistant Professor at the Department of Communication and Media, Lund University, Sweden. His research concerns veganism and nonhuman animals as food, in particular focusing on the dairy industry. Together with Helena Pedersen he developed the course *Critical Animal Studies: Animals in Society, Culture and the Media* at Lund University. The course was awarded the "Distinguished New Animals and Society Course Award" by the Humane Society of the United States in 2012. In 2013–2014 Tobias Linné served as coordinator (together with Helena Pedersen and Amelie Björck) for the research theme "Exploring 'the Animal Turn': Changing perspectives on human-animal relations in science, society and culture", funded by the Pufendorf Institute for Advanced Studies at Lund University. In 2014, he was an invited research fellow of the New Zealand Centre for Human-Animal Studies in Christchurch, New Zealand.

Jordan McKenzie
completed his PhD at Flinders University in South Australia and is currently a lecturer in sociology at the University of New England, NSW. His interests are grounded in social and critical theory with a specific focus on theories of emotion, participatory democracy and reason. In particular, his work engages with theories of social experience formed through critiques of modernity and modernization, and this can be found in his current research on happiness and contentment.

Helena Pedersen
is Associate Professor of Education at the Department of Child and Youth Studies, Stockholm University. Her primary research interests include Critical Animal Studies, critical theory, educational philosophy and posthumanism. She is author of *Animals in Schools: Processes and Strategies in Human-Animal Education* (Purdue University Press, 2010), which received the Critical Animal Studies Book of the Year Award in 2010. Helena Pedersen is also co-editor of the Critical Animal Studies book series (Rodopi/Brill) and serves on the editorial board of *Other Education: The Journal of Educational Alternatives*. Together with Tobias Linné and Amelie Björck she coordinated the research theme "Exploring 'the Animal Turn': Changing perspectives on human-animal relations in science, society and culture", funded by the Pufendorf Institute for Advanced Studies at Lund University 2013–14.

Annie Potts
is an Associate Professor in Cultural Studies and Human-Animal Studies at the University of Canterbury, where, along with Philip Armstrong, she also directs

the New Zealand Centre for Human-Animal Studies (www.nzchas.canterbury.ac.nz). Annie is the author of *The Science/Fiction of Sex: Feminist Deconstruction and the Vocabularies of Heterosex* (Routledge, 2002) and *Chicken* (Reaktion 2012); co-author (with Philip Armstrong and Deidre Brown) of *A New Zealand Book of Beasts: Animals in our Culture, History and Everyday Life* (Auckland University Press 2013) and (with Donelle Gadenne) of *Animals in Emergencies: Learning from the Christchurch Earthquakes* (Canterbury University Press 2014); and the editor of a special issue of the journal *Feminism & Psychology* on 'Feminism, Psychology and Nonhuman Animals'. She has served on the National Animal Welfare in Emergencies Management Advisory group, the New Zealand Companion Animal Council, and First Strike New Zealand. In 2014, she received a New Zealand Assisi Award for Services to Animal Welfare.

Vasile Stănescu
received his Ph.D. from Stanford University in the program of Modern Thought and Literature. He is the Co-Senior Editor of the *Critical Animal Studies* Book Series published by Brill/Rodopi Press. Vasile serves as Assistant Professor and Director of the Program in Speech and Debate at Mercer University. He is currently working on publishing a book entitled *Happy Meals: Animals, Nature, and the Myth of Consent*.

Kate Stewart
is Principal Lecturer in Sociology at Nottingham Trent University in the United Kingdom. For the past decade she has had a particular research interest in how information about food is interpreted and applied. This research led to her first collaborative work with Matthew Cole, "The Conceptual Separation of Food and Animals in Childhood," published in *Food, Culture and Society* in 2009.

Nik Taylor
is an Associate Professor at Flinders University in Adelaide, Australia. As a sociologist she has been researching human-animal relations for over 15 years, after spending years running an animal shelter. Nik has published 6 books and over 40 journal articles and book chapters on the human-pet bond; treatment of animals and animal welfare; links between human aggression and animal cruelty including those between domestic violence, animal abuse and child abuse; slaughterhouses; meat-eating, and, animal shelter work. She has written for diverse audiences including *The Guardian, The Drum, The Conversation* as well as numerous blogs and websites. Her most recent books include *The Rise of Critical Animal Studies* (edited with Richard Twine, Routledge, 2014), *Humans, Animals and Society* (Lantern Books, 2013) and *Animals at Work* (with Lindsay Hamilton, Brill Academic, 2013).

Richard Twine

is a Senior Lecturer in Social Sciences and is the Co-Director of the Centre for Human Animal Studies at Edge Hill University in the United Kingdom. His current research falls under food transitions and the Sociology of Climate Change with a strong concurrent interest in gender studies, ecofeminism, veganism, environmental social science, and critical animal studies. Richard is the author of *Animals as Biotechnology: Ethics, Sustainability and Critical Animal Studies* (Routledge, 2010) and co-editor (with Nik Taylor) of *The Rise of Critical Animal Studies: From the Margins to the Centre* (Routledge, 2014).

Dinesh Joseph Wadiwel

is a Lecturer in Human Rights and Socio-Legal Studies at The University of Sydney. His research interests include sovereignty and the nature of rights, violence, race and critical animal studies. He is author of *The War Against Animals* (Brill, 2015).

Yvette Watt

is a Lecturer in Fine Art at the Tasmanian College of the Arts, University of Tasmania, where she also completed a MFA and a PhD. She is a committee member of Minding Animals Australia and Co-Director of the UTAS Faculty of the Arts Environment Research Group. Yvette's art practice spans 30 years. She has held numerous solo exhibitions and has been the recipient of a number of grants and awards. Her work is held in numerous public and private collections including Parliament House, Canberra, Artbank and the Art Gallery of WA. Yvette has been actively involved in animal advocacy since the mid-1980s, including being a founder of Against Animal Cruelty Tasmania, and her artwork is heavily informed by her activism. She is a contributor to and co-editor (along with Carol Freeman and Elizabeth Leane) of the collection titled *Considering Animals: Contemporary Studies in Human-Animal Relations* (Ashgate, 2011). Other essays by Yvette include 'Artists, Animals and Ethics' in *Antennae* (2011, issue 19) and 'Animal Factories: Exposing Sites of Capture' in *Captured: Animals Within Culture* (Palsgrave McMillian, 2014, edited by Melissa Boyde).

CHAPTER 1

What is Meat Culture?

Annie Potts

> The exploitation of animals for profit is enabled by a cold, calculating Trinity of Science, Technology, and the Market that has stripped our public life of empathy.
> KIMBRELL 2010, 29

⋮

More people eat meat than ever before, and global meat consumption continues to grow. This is not simply an effect of the rising human population (7.3 billion as of January 2016):[1] it is the result of rapid scientific, technological and sociocultural changes that have transformed meat production and consumption over the past one hundred or so years, and especially in industrialized countries following World War II. The history of creating contemporary 21st century 'meat culture' is an industrial history that blends agricultural science and technologization with mass production, vertical integration production systems with globalized economies, and the hyper-stimulation of consumer demand emblematized by the rise of suburban fast food outlets since the 1950s. Colossal shifts have occurred over the last half century with respect to the breeding, farming, slaughter and consumption of animals. Genetic engineering and selective breeding have created new 'hybrid' animals who confound nature by eating less but growing fatter faster. Smaller family farms have been replaced by massive Concentrated Animal Feeding Operations (CAFOs or factory farms), where animals are incarcerated their entire lives in cramped cages, pens or sheds and subjected to extreme physical, mental, social and emotional stress. Slaughtering processes have been sped up to cope with the vast numbers of animals being killed each year to meet the demand for animal flesh. At the start of the 20th century the city of Chicago—known as "the cradle of the slaughter industry"—was killing up to 12 million animals annually (*Meat Atlas* 2014, 14). Today, in just one of the meat industry's key companies,

---

[1] See http://www.worldometers.info/world-population/. Accessed 14 January 2016.

Tyson Foods, over 42 million chickens, 170,000 cattle and 350,000 pigs are killed *each week*, these animals coming from the company's own CAFOs (ibid).

The analysis of meat and its place in Western culture has been central to Human-Animal Studies as a field. Texts published in the early 1990s problematizing the hegemony of meat-eating—such as Carol Adams' *The Sexual Politics of Meat: A Feminist-Vegetarian Critical Theory* and Nick Fiddes' *Meat: A Natural Symbol*—were pivotal in establishing and growing this new cross-disciplinary area of study. In the 25 or so years since these first publications emerged, scholars in the humanities and social sciences have continued to interrogate the various representations, meanings, practices, ethics, and modes of identity associated with meat production and consumption (and also its opposite, veg*nism).[2] Attention has been directed to issues such as meat's portrayal in popular culture, including meat (and dairy) industry marketing and advertising (Adams 2003, Packwood Freeman 2009, Cole 2011, Pilgrim 2013, Taylor 2016); the gendered construction of meat consumption (and of animal slaughter) (Adams 2010, Luke 2007, Parry 2010, Potts & Parry 2010); the shifting technologies and capitalist economies connected to meat production, distribution and procurement (Noske 1989, Horowitz 2006, Marcus 2005, Twine 2010); the politics and ethics of selective breeding and genetic modification of 'farmed animals', including the killing of infants born into but 'surplus' to the meat or dairy industries (Imhoff 2010); the suffering of animals contained in Concentrated Animal Feed Operations (CAFOS), as well as those born into free range farming situations (Eisnitz 2007, Foer 2010, Joy 2010, Lappe 2010, Pachirat 2011); and the environmental impacts of intensive farming (Twine 2010, Taylor 2012).

This volume builds on and advances the existing critical examination of meat's place in Western societies, bringing into urgent focus a wide range of domains of production and consumption of animals within the coherent framework of what we have chosen to call 'meat culture'. Featuring new work from key Australasian, European and North American scholars, each chapter interrogates in depth some aspect of the animal industrial complex (Noske 1989) and meat hegemony in the 21st century.

This first chapter presents an overview of the global meat industry, drawing attention to the actual lives and deaths of the animals that are integral to, yet routinely obscured by, meat industry statistics, narratives about farming, and economic rhetoric. It also introduces the concept of 'meat culture', and describes how each chapter comprising this volume scrutinizes a distinct manifestation of carnist ideology.

---

2 The term 'veg*n' is commonly used to connote both vegetarians and vegans.

Making Meat, Counting the Costs

In 2011, 296 million cattle, 24 million buffalo, 1.383 billion pigs, 430 million goats, 517 million sheep, 654 million turkeys, 2.8 million ducks, 649 million geese and guinea fowl, and a staggering 58 billion chickens were killed across the world (*Meat Atlas* 2014). Over the past fifty years, global meat production from land animals (including birds) has almost quadrupled from 78 million tonnes in 1963 to 308 million tonnes in 2013 (and a forecasted 319 tonnes in 2015).[3] By 2050 it is estimated to reach 455 million tonnes.[4]

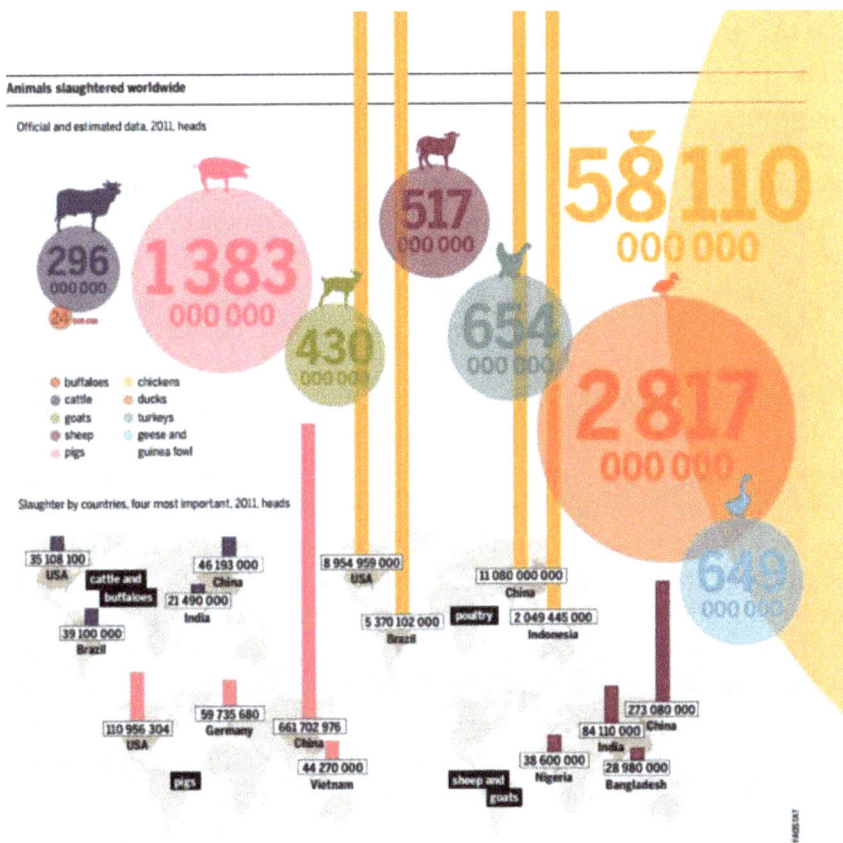

FIGURE 1 *Numbers of animals slaughtered in 2011.*
COURTESY OF *MEAT ATLAS*.

---

3 See http://www.globalagriculture.org/report-topics/meat-and-animal-feed.html. Accessed 14 January 2016.
4 See http://www.fao.org/3/a-i4136e.pdf. Accessed 14 January 2016.

Of the 308 million tonnes of meat 'produced' in 2013, approximately 114 million tonnes came from pigs, 106 million tonnes from birds (mostly from chickens), 68 million tonnes from bovines (consumed as beef or veal), and 13.8 tonnes from sheep and goat flesh. Meanwhile, as many as *2.7 trillion* marine species are estimated to be killed annually via commercial fishing operations (Mood and Brooke 2010). This figure does not take into account those killed in aquaculture (fish farms).

In the United States, the nation that propagated and most avidly advanced the intensive farming practices now commonly utilized by Western nations, we can trace changing patterns of meat consumption over a fifty year period. In 1960, 1.8 billion animals were slaughtered in the USA: by 2013 this number had risen to 9.1 billion, the vast majority of those killed for meat were birds (Humane Society of the United States Farm Animal Statistic: Slaughter Totals 2013). In 1960, the number of cattle and buffaloes slaughtered for meat was over 25 million; by 2013 this number had risen to 32.5 million. In 1960 79 million pigs were slaughtered; by 2013 112 million pigs were killed annually. In 1960 1.6 billion chickens were slaughtered for food in the United States, but by 2013 this number had catapulted to nearly *9 billion* (an increase of 600%). The fate of turkeys is as shocking: nearly 71 million turkeys were slaughtered in the USA in 1960: by 2013 this number had increased to 239.4 million (*Meat Atlas* 2014).

When taken as a whole, figures supporting the increase of global meat consumption conceal how growth is occurring in some regions while slowing down in others. In China and India an 80% rise in meat consumption is anticipated by 2022. These nations are undergoing substantial economic growth, which, when coupled with the effects of globalization—including the introduction of new Westernized foodstuffs—has initiated a shift (predominantly in the wealthier urban classes) from plant-based meals to meals higher in meat and dairy.[5] Similarly, in the African continent, where meat has not previously been a staple of the typical diet (the average African consuming around 20kg of animal flesh annually compared with the average American's consumption of around 118kg),[6] certain countries such as South Africa, Egypt, Nigeria, Morocco and Ethiopia, are now investing more in animal farming and meat consumption is also quickly increasing.

---

5   Globalization works in all directions of course, so while Japanese people are now eating steak, Hindu Indians consuming burgers, and Chinese dairy products, more and more Westerners are drinking soy milk and eating tofu.
6   http://www.globalagriculture.org/report-topics/meat-and-animal-feed.html. Accessed 15 June 2015.

In contrast, there is evidence that in those nations where the mass production of meat via intensive farming first occurred, the consumption of animals is slowing down if not stagnating. That is, the increasing demand for meat in developing countries is occurring alongside a decline in meat consumption in industrialized nations. For example, meat consumption in the USA dropped 9% from 2007–2012. The decline in meat eating in the Western world has been attributed to growing awareness of the health risks associated with eating red meat (Pan et al 2011), increased concern about the environmental impacts of animal farming (including the extensive crop farming required to feed animals) (Twine 2010), and rising ethical concerns about the welfare of animals in intensive farming systems and slaughterhouses (Marcus 2005; Eisnitz 2007). This decline is noticeable, however, only with respect to certain species of farmed animals. For example, the consumption of meat from cattle and sheep has plateaued or fallen, while consumption of meat from pigs and chickens continues to rise. In part this has been driven by the greater expense of red meats (compared to so-called 'white meats'), and in part by concerns about the detrimental effects of red meat on human health (the National Institutes of Health reporting in 2012 on "evidence that eating red meat on a regular basis may shorten your lifespan").[7] By 2020 it is predicted that over 124 million tonnes of poultry and 120 million tonnes of pig flesh will be produced globally—an increase of 25% in just ten years (*Meat Atlas* 2014).

Undeniably the species *most* affected by global demand for meat is the chicken. By 2020, China's production of meat from chickens will increase by 37%, Brazil's by 28%, and the USA's production by 16%. In India, poultry consumption is expected to rise nearly tenfold to around 10 million tonnes per year by 2050. Chickens are easier and more lucrative to farm because they are cheaper to feed and house than other animals: they can be confined in the millions on one factory farm where human labour costs are minimal as feeding and watering are automated via machines. There are few religious taboos restricting consumption of bird flesh so chickens are more acceptable than other species on menus around the world (and therefore more profitable). Australians eat the most chicken meat (50.5 kg per person per year), followed by Americans (50.1 kg) and Brazilians (38.5 kg). Those living in the European Union consume on average 23.6 kg of chicken meat per year, those in China 14kg, and those in India 2.4 kg (*Meat Atlas* 2014).

---

7   See http://newsarchive.medill.northwestern.edu/chicago/news-218052.html.

The Suffering Behind Modern Meat Production

All these numbers and percentages and trends obscure the fact that behind each statistic are *trillions of individual mammals, birds and marine creatures* captured and/or confined in various farming systems until their lives are prematurely and violently terminated by a blade, a macerator, gas, scalding water, a hook, a net or some other instrument. It is beyond the scope of this introductory chapter to discuss the plight of all animals farmed and eaten on the planet (the crocodiles and alligators, frogs, snails, eels, octopuses, crabs, sharks, shellfish, eels, ducks, geese, emu, rabbits, guinea pigs, horses, deer, sheep, goats, dogs, cats, kangaroo, to name just a few), but the following (albeit very brief) snapshots focus on some of those exploited in the greatest numbers.

Pigs

Eurasian wild pigs, from which domestic pigs have descended, are curious and sociable mammals inhabiting open forests and woodlands in matriarchal herds comprising half a dozen closely related females and their offspring (sub-adult males sometimes form herds while mature boars are solitary). Herds roam over about 25 hectares, foraging for food with their bony, muscular snouts, constantly communicating with each other through squeals, grunts and chirps. At night the herd sleeps together in large dens or nests. Pigs love wallowing in mud, an activity which cools them in summer and removes parasites; they are naturally clean animals and leave their waste in specific toilet areas (Serpell 1996). Domestication of pigs occurred 9000 years ago but none of these behaviours and desires have been lost to those pigs confined today in the cages and pens of the pig meat industry. When researchers at the university of Edinburgh permitted domestic pigs to run in a 'pig park' approximating their natural habitat in the wild, the sows foraged for a mile during the day, carefully built nests on a hillside so urine and faeces flowed downhill, and took turns minding the piglets.[8] Modern consumers are generally ignorant of the natural behaviours, characteristics and pleasures of pigs, more likely to assume pigs are dirty and greedy, and only 'know' pigs as "the 18 per cent ham, 16 per cent bacon, 15 per cent loin, 12 per cent fatback, 10 per cent lard and 3 per cent each of spare rib, plate, jowl, foot and trimmings that exit the modern packing plant" (Mizelle 2011, 7).

The pigs born to intensive farming could live up to fifteen years but most are slaughtered at just six months of age, having been removed from their mothers within 10–21 days of birth (infant males are castrated to improve the taste

---

8 See www.ign-nutztierhaltung.org. Accessed 14 June 2015.

of their flesh for consumers) and 'grown' to 'slaughter weight' in small overcrowded pens (usually the size of a small room but containing up to 20 pigs) kept in immense windowless sheds referred to as 'finishing sheds' (Marcus 2005). They have no experience of sunlight or of the outdoors (until they are transported to the slaughterhouse), and their heightened sense of smell, which in the wild assists them to navigate and to root out food underground, is assaulted by the inescapable ammonia fumes amassed from their own waste. Sows in factory farms spend most of their reproductive lives in cages: when pregnant they are kept in cages (euphemistically referred to as 'gestation stalls' or crates), typically 7 feet long and 2 feet wide, which do not permit them to even turn around. Because their lives are so constrained, diminished and frustrating, sows perform stress-related behaviours such as repetitively chewing on the bars of these cages. When nursing their piglets they are moved to farrowing cages; after the baby pigs are taken away from them sows are inseminated again, and this cycle continues until, no longer deemed to be productive, they are dispatched to the same slaughterhouses that have already killed their offspring.[9]

Cattle

Cattle are members of the *Bovidae* family, which includes 140 species including goats, sheep, antelopes, bison and buffalo. They live in close-knit groups of up to twenty, roaming grasslands, scrublands, forests and even deserts. The social structure of the herd is matriarchal with mothers and daughters forming extremely close bonds and staying together for life (Young 2009). Cattle bred for the beef industry may spend some time as part of a herd grazing on grassland, but in countries where CAFOs are prominent, calves are at around six months of age transferred inside to 'feedlots'—factories where up to 100, 000 cattle can be housed at once. Here their food intake is increased and movement limited in order to ensure they put on the greatest weight in the shortest time. In the USA, 90% of these CAFOs also add growth hormone to food in order to improve weight gain, while 83% administer antibiotics to prevent animals becoming ill as a result of their unnatural confinement. While technically still infants, at around 12–16 months of age, these unnaturally fattened cows and steers are slaughtered.

Around 378 billion litres of milk is processed each year by dairy companies around the world with global demand for dairy products continuing to grow

---

[9] ASPCA, https://www.aspca.org/fight-cruelty/farm-animal-cruelty/pigs-factory-farms. Accessed 10 June 2015.

(Fonterra 2015). In 2015 this upbeat promotion targeting New Zealand dairy farmers was announced:

> Blue Sky Meats has developed some new and exciting initiatives for all of our bobby calf suppliers for the coming processing season. All suppliers will go into a draw to win a trip for two to the All Blacks vs. England game[10] in London in November [...] to be topped off with a visit to Paris and a specially organised gourmet meal, dining on Blue Sky's processed bobby veal product at one of our French customer's restaurants. Every calf supplied entitles the supplier to go into the draw to win this incredible trip. So the more calves supplied the greater chance you have of winning! (Blueskymeats 2014).[11]

The Blueskymeats campaign demonstrates how the dairy industry is always as much about killing as it is about milk production (in fact 55% of beef consumed in New Zealand comes from the dairy industry). In the year this advertisement appeared, New Zealand, the world's top milk exporter per capita (a nation with 4.7 million humans, 6.7 million cattle and 29.6 million sheep) killed 2 million baby male calves born to dairy cows. The milk lactated by these cows and meant for their calves went instead to humans. Under natural conditions calves stay with their mothers for eight to eleven months: the majority of these male dairy calves, not considered the 'right breed' to make beef and unable to produce milk, are immediately separated from their mothers and sent to slaughter as 'waste products' of the dairy industry. Nor do female calves escape this fate—those not wanted to replace their mothers in the dairy sheds are also killed soon after birth. In New Zealand, these 'bobby calves', as they are called, may legally be denied food for up to thirty hours before being collected by stock trucks and taken to 'freezing works' (as slaughterhouses are euphemistically referred to in New Zealand). For male calves born to the dairy industry and not immediately killed, this abrupt and unnatural separation from their mothers turns into confinement on chains within cages, fed mainly on milk with few or no solid foods (a diet designed to deprive muscle growth, and which will also lead to anaemia) in order to produce paler calf flesh, renamed for sale as 'veal', and considered a delicacy by the 'discerning'

---

10 This refers to rugby, New Zealand's 'national sport'.
11 See http://bluesky.co.nz/Supply/Supplier/Bobby-Calves. Accessed 10 June 2015.

FIGURE 1.2   *Advertisement placed by New Zealand animal rights organization Save Animals From Exploitation (SAFE) in* The Guardian *newspaper in late 2015. This was part of an effective campaign aiming to raise awareness about the fate of male calves born into the dairy industry.*
COURTESY OF SAFE (www.safe.org.nz).

middle class consumer. Six million calves are raised for veal in the European Union per year, the highest producers being France, the Netherlands and Italy.[12]

The process of developing our modern day milk factories began around two centuries ago when demand for milk propelled dairy farming on the path towards industrial-scale production where the "biological limits of lactating mothers are treated as inconvenient obstacles" (Mendelson 2010, 131). The average per capita global milk consumption amounts to about 100kg of milk per year, with significant differences across countries and regions. Per capita consumption in Western Europe, for example, is in excess of 300 kg milk per year compared with 10–30kg in some African and Asian countries (Mendelson 2010). From 2002–2007 world milk production rose by 13% to 697 million tonnes per annum, with China, India and Pakistan alone accounting for two thirds of all volume growth. The natural lifespan of a cow is over twenty years (see chapter seven in this volume); however, as Mendelson (2010, 131) points out, "the farms supplying milk for breakfast cereal are likely to send most members of the herd off to the hamburger plant within three or four years of their first lactation".

Fish

> Winston Churchill called them 'the good companions'. John Lennon smothered his in tomato ketchup. Michael Jackson liked them with mushy peas. They sustained moral through two world wars and helped fuel Britain's industrial prime.

So writes BBC News journalist James Alexander when reporting on the 150th anniversary of Britain's 'national institution': fish and chips. Today in Britain, burgers, fried chicken, pizza, Indian and Chinese takeaways are more popular than fish and chips (this is reflected in the reduced number of 'chippies'—from 25,000 in 1910 in the UK to 10,000 in 2009), however 'fish and chips' remains a recognized family treat in the United Kingdom (as well as in its former colonial outposts such as New Zealand and Australia), where this meal represents family time, the end of a working week, a night off cooking, holidays and relaxation. The benign symbolism of this traditional working class food belies the suffering of countless individual fish and other marine life killed for consumption or caught as casualties of the commercial fishing industry and in aquaculture (fish farms).

---

12 See http://www.ciwf.org.uk/farm-animals/cows/veal-calves/. Accessed 13 June 2015.

As many as 2.7 trillion marine animals are taken from the seas and oceans each year, while the contribution of intensive fish farming to total fish 'production' is increasing at an average annual growth rate of 6.1% (from 36.8 million tonnes in 2002 to 66.6 million tonnes in 2012), with Asia producing more farmed fish than wild catch in 2008.[13] 'Aquaculture' may sound benign but it is another version of intensive farming. Like broiler chicks and layer hens in battery farms, pigs in 'crates', and cattle in feedlots, the fish in these systems endure life in packed pools. Such high stocking density leads to injuries as fish try to behave as they would in nature; and, to make matters worse, farmed fish are typically starved for a week before slaughter in order to empty their guts (Stone 2011).

The global slaughter of fish, crustaceans, molluscs and other aquatic animals continues to increase, having reached 158 million tonnes of marine meat in 2012. The world fishing fleet comprises around 4.7 million vessels—the nations most involved in commercial fishing are China, Indonesia, the United States, India and Peru. Worldwide, anchoveta (a member of the anchovy family) is the species killed most, followed by Alaska Pollock, skipjack tuna, Atlantic herring and chub mackerel. Eighty six percent of fish killed are consumed directly, while 14% are converted into fishmeal and fishoil (46% of fish destined for human consumption are live and in fresh form) (Food and Agriculture Organization of the United Nations Fisheries and Agriculture Department 2015).

The flesh from fish and other marine creatures may not always be understood as 'meat' but it should be; this categorical error is largely generated and perpetuated by Judeo-Christian beliefs about the difference between the value and consumption of creatures of the sea versus land-based animals (demonstrated, for example, in the practice by the devout of 'replacing' (animal) meat with fish on religiously-determined 'meat-free' days). Attitudes in Western nations are influenced by anthropocentric discourses that position humans as superior to other species in a hierarchy where creatures deemed the most unlike humans (in physical appearance or form, ways of perceiving and experiencing the world, and/or modes of living) are deemed of least value or worth. Fish are relegated near to or at the bottom of this hierarchy. They are viewed as so different to humans that they have commonly been disregarded as sentient creatures. For those who need proof of fish sentience, there are numerous recent studies now showing how fish suffer from pain and distress (Braithwaite

---

13   These figures associated with fish farming break down to 44.2 million tonnes of finfish, 15.2 million tonnes of molluscs, 6.4 million tonnes of crustaceans, and 0.9 million tonnes of other aquatic animal species (Food and Agriculture Organization of the United Nations Fisheries and Agriculture Department 2015).

2010, Balcombe 2016). This fact is extremely sobering in light of the ways in which fish are killed. Most commercially caught fish are alive when brought out of the water: they are distressed, injured and in pain, and they die slowly either from asphyxiation or from a combination of asphyxiation and live gutting. The time taken for sensibility to be lost varies according to the method of killing. Fish live for 20–60 minutes when killed via asphyxiation and live gutting; they can live between one and four hours when killed through asphyxiation alone. Live chilling prolongs death. Fish caught by trawl nets also suffer from exhaustion and can die from the weight of other fish on top of them, or as a result of decompression effects (burst swim bladders). Fish caught in nets or via hooks on long lines may be alive for days after capture (Vis and Kestin 1996, Robb and Kestin 2002).[14]

Chickens

> Question: Why did the chicken go to KFC? Answer: He wanted to see a chicken strip.
> Question: Which day of the week do chickens hate most? Answer: Fry-day!
> Question: How did the headless chicken cross the road? Answer: In a KFC bucket.[15]

As these jokes attest, the chicken is a much trivialized and belittled bird. Chickens are also the species whose flesh is consumed the most across the world, as the statistics presented earlier demonstrate. Renowned neuroscientist Lesley Rogers, author of *Brain and Development in the Chicken* (1995), argues that we owe it to chickens to understand them much better as their species is the one humankind has singled out to manipulate and exploit in the greatest numbers. Tragically, however, chickens continue to be so objectified and over-determined as 'food' that they are likely to be viewed as meat rather than living beings *when still alive*.

The rapid and tragic ascendancy of the factory farming of chickens for meat and eggs began with the development in the late 1800s of two new farming technologies—the incubator and colony brooder—which essentially removed

---

14   While this volume does not include further analysis on the farming, capture, slaughter and consumption of fish and other marine life, the reader is referred to Victoria Braithwaite's (2007) excellent *Do Fish Feel Pain?* and Jonathan Balcombe's forthcoming book, *What a Fish Knows*, due for publication in 2016.
15   Sourced from: http://www.jokes4us.com/animaljokes/chickenjokes.html, 10 June 2015.

the previous seasonal limitations on chicken farmers. In nature a hen will lay a clutch of eggs each spring and raise these chicks over the course of several weeks. But with the advent of machinery that could replace mother hens, the process of separating chicks from hens and from the natural environment, commenced in earnest. During the first part of the 20th century scientists examined all aspects of chicken histology, physiology, anatomy, genetics, nutrition and pathology—every biological feature was experimented on with the aim of rendering chickens more serviceable to humans. The 'artificial evolution' of chickens through selective breeding also resulted in the separation of egg farming from the farming of chickens (or, in fact, baby chicks) for meat. However the most profound change was the confinement of thousands of chickens indoors for the purposes of management and control.

The global broiler or meat chick of today is a combination of Cornish male and White Plymouth Rock female lines. Such breeds, referred to as 'industrial poultry stocks', are given numbers not names in order to control commercial information and dispersion. In 1920 broiler chickens averaged 1 kg (2.2 lb) when killed at around sixteen weeks of age; by 1941 they weighed 1.3 kg (2.9 lb) when slaughtered at twelve weeks; today, chicks average 2.7 kg (5.9 lb) when killed at six weeks of age. Daily rates of growth have increased by 300% (from 25g or 0.88 oz to 100 g or 3.52 oz) over fifty years. Moreover the amount of food required to bring a broiler chick to slaughter weight today is radically less: seventy years ago 6.5 lb (3 kg) of feed was required to produce 1 lb (0.5 kg) of chicken meat; now it takes a mere 1.75 lb (0.8 kg). Scientists have therefore created a bird who, by eating less, grows faster (Potts 2012).

In nature, chickens are gregarious, curious and busy. They have acute hearing, sensitive taste, and extraordinary panoramic and colour vision, which enables them to peck at food on the ground while also keeping an eye on possible predators approaching by sky. They live in small flocks comprised of a rooster and several hens, and follow special hierarchical social structures which make them feel safe and secure. Chickens form friendships with each other and may withdraw from the flock to grieve when friends die (Potts 2012; see also chapter ten in this volume). The strong bond between a mother hen and her chicks is well known and represented in stories and idioms across human cultures.

The life of a chicken born to the commercial meat industry is short and miserable. Confined in overcrowded stinking sheds, unable to fulfil any natural behaviours (including some of the most crucial such as establishing and maintaining flock social etiquette), and completely without the pleasures of sunshine, grass, dirt to scratch in and trees to roost in, these birds are viewed as 'objects' and 'things' by those who grow them, and by those who consume them. As a consequence of such intense and unrelenting confinement, and because their bodies are growing too large to support their skeletons and organs, chicks develop physical and emotional conditions. They may suffer heart failure, have trouble breathing and walking, and become unable to stand. When crippled like this they may starve or die of thirst because they are unable to push through the other birds to reach the automatic food and water dispensing machines. The emotional lives of broiler chicks are just as distorted and restricted as their physical forms. They may become so overwrought by the compromised state of their bodies and their overcrowded, relentlessly noisy and filthy environment that any unexpected sound or movement can cause a mass panic, a phenomenon termed 'broiler hysteria' within the industry (Potts 2012). In order to ensure they grow as big as possible, chicks raised for meat are also kept in almost constant light so they don't sleep and remain stimulated to eat. Between four and six weeks of age, while still technically baby birds these chicks are caught during darkness and transported to the slaughterhouse.

America, where the broiler meat industry began in the 1920s in Delamare, and where KFC and McNuggets originated, continues to idolize fried chicken, but today the greatest activity in broiler chicken production is occurring in Asia and South America. In China in the early 1980s chickens were typically kept in small flocks in backyards; by 1997, over 63,000 concentrated animal feeding operations (known as CAFOs) had emerged in this nation, with single farms housing up to 10 million birds (Greger 2009). Indeed, consumption of chicken meat in China is estimated to have risen by 50% over the decade between 2000 and 2010 (substantial increases in broiler meat consumption have also occurred in Brazil, Mexico, South Korea, Taiwan, Indonesia and the Philippines) (Liying et al 2001).

Of course chickens bred for the egg industry also end up on plates. Like the dairy industry, the business of egg farming has no use for male chicks (except for those kept for breeding), and so around 50% of all chicks are, within 24 hours of hatching, fed live into a large high-speed grinder, a process of slaughter known as 'instantaneous maceration'.[16] And, as is the case for sows

---

16 Germany leads the way in outlawing the practice of shredding alive newly hatched male chicks. After pressure from animal activists the German government cooperated with scientists to develop an alternative method of testing for and removing male embryos from hatcheries. New technology will determine the sex of embryos before chicks hatch; all male-identified embryos will be removed from hatcheries, and the eggs they grew in will be used for other products. See: http://metro.co.uk/2015/10/24/germany-becomes-the-first-country-to-ban-disturbing-chick-shredding-practice-from-egg-industry-5459759/#ixzz3xH8qWoMw. This new technology is to be established in all egg farms in Germany by the end of 2016.

FIGURE 1.3 *Because they have been bred specifically to grow fatter quicker, these four week old broiler chicks, raised for their meat (not for eggs), are already unable to stand up. These birds, rescued by Chocowinity Chicken Sanctuary, North Carolina, did not survive long after this photo was taken.*

and cows, when hens' reproductive lives are over and they stop producing eggs (exhausted after 18 months to two years), these birds are killed and consumed in soups, baby foods and stock cubes (Potts 2012).

Humans

The same numbers and percentages that obscure the actual lives and deaths of those animals raised and killed for the food industry also conceal the impact on humans working at the frontline of slaughter, dismemberment and meat packaging. In the United States and elsewhere cheap meat is produced through cheap labour. Immigrant and unskilled workers are exploited in the production of meat—denied union power, real wages, health care benefits, and even bathroom breaks (Cook 2010). In the chicken meat industry, whether catching, hanging or eviscerating birds, the processing of around nine billion chickens per year in order to meet the American demand for fast food means employees are required to work on high-speed assembly lines, often succumbing to occupational overuse conditions such as carpel tunnel syndrome, and incurring disabling injuries when hit by fast powered machinery used to maximise efficiency and profits (Pachirat 2011). The typical chicken slaughterhouse in the United States can, within an eight hour shift, convert 144,000 live birds into thousands of packages of 'ready-to-cook chicken' (Cook 2010, 236).

In New Zealand, where animal farming is popularly framed as 'the backbone of the nation's economy' (Potts, Armstrong and Brown 2013), the expansion of dairy farming has been coupled with a rapid increase in the number of migrant (particularly Filipino) workers recruited to dairy farms, usually on temporary work visas (Fulton 2013). More detailed stories have emerged in the media in the past year or so of workers' exploitation, including dishonest immigration recruitment strategies, 18-hour working days (with stretches of eight days at a time), unfair minimal wages, lax health and safety training and monitoring, and even physical and sexual abuse in the work environment (Meadows & Cronshaw 2015).

The Planet

> For each hamburger produced from animals raised on rainforest land, approximately 55 square feet of forest have been destroyed.
> STONE 2011, 37

∙ ∙ ∙

> A farm with 2,500 dairy cows produces the same amount of waste as a city of 411,000 people.
>
> US Environmental Protection Agency 2004

∴

The farming of animals to produce meat has well-known deleterious effects on the environment.[17] Intensive farming results, for instance, in 18% of greenhouse gas emissions measures in CO_2e (carbon dioxide equivalent), more than all global emissions from transport including airplanes, trains and automobiles (Twine 2010, 135). Agribusiness accounts for 37% of methane emissions, 65% of nitrous oxide emissions, and 64% of human-induced ammonia emissions (Stone 2010, 37).

Thirty three percent of arable land in the world is currently dedicated to raising crops to feed animals, and an additional 36% of the ice-free terrestrial surface of the planet is used for grazing animals (Lappe 2010). In the Amazon over 70% of deforested land is exploited for grazing: a land area equal to seven football fields is destroyed in the Amazon basin every minute. Farming of animals also accounts for 55% of soil erosion (Stone 2011). The excessive use of chemical fertilizers, pesticides and herbicides in the production of crops to feed animals harms organisms in the soil and water, and damages ecosystems. Farming uses vast quantities of water. In the United States, 55 % of water is consumed by the agricultural industry, compared to 5% by private households. 2,500 gallons of water are required to make one pound of beef.

Worse is to come with global meat production projected to double from 229 million tonnes in 1990 to 465 million tonnes by 2050, and milk production to increase from 580 tonnes (1990) to 1043 tonnes (2050) (FAO, *Livestock's Long Shadow* 2006). Unfortunately these impacts on the environment are taken less seriously by governments and producers in established and growing capitalist economies because they cannot be easily or precisely measured in economic or monetary terms (although the European Nitrogen Assessment in 2011 estimated this damage amounted to 70–320 billion euros in Europe, a figure which exceeds all the profits made by the continent's agricultural sector) (*Meat Atlas* 2014).

---

17 See *Livestock's Long Shadow*, accessed via ftp://ftp.fao.org/docrep/fao/010/a0701e/a0701e.pdf.

Contemporary 'Meat Culture'

There is ample evidence verifying meat consumption is harmful to humans' health[18] and to the planet, and there is also more than enough evidence to show that nonhuman species farmed for meat are sentient beings fearing death[19] and suffering greatly as a result of modern farming practices. And yet meat retains a central place on the table in many homes throughout industrialized nations, and is increasingly common in developing countries too. It has recently been gauged by Vegetarian Calculator that the average omnivore in industrialized countries will eat around 7,500 animals in a lifetime of eighty years. This figure breaks down to 11 cows, 27 pigs, 30 sheep, eighty turkeys and 2,400 chickens, 4,500 fish and unspecified numbers of goats, rabbits, squid, prawns, and other 'seafood' (Pleasance 2015).[20]

What propels this investment in meat? And what sustains the belief in animal flesh as an acceptable food in the 21st century?

Clearly technological changes have enabled the selective breeding, intensive farming, speedy killing, dismembering and packaging of nonhuman species, but meat is not just the product of these changes. It is also a 'food' that is heavily infused and invested with meaning. In all human cultures it is also symbolic: in the Western context it signifies important ideas about gender (Adams 2010, Parry 2010, Potts and Parry 2010, Hovorka 2012), class and taste (Potts and White 2008), socioeconomic position (Galobardes et al 2001), geographical and economic factors (Hovorka 2008). Its acceptance is facilitated by beliefs about

---

18   See, for example, Michael Moseley (2014) "How Safe is Eating Meat?" BBC, http://www.bbc.co.uk/news/health-28797106; and Ian Sample (2014) "Diets High in Meat, Eggs and Dairy Could be as Harmful to Health as Smoking", *The Guardian*, http://www.theguardian.com/science/2014/mar/04/animal-protein-diets-smoking-meat-eggs-dairy; Animal Protein Diets http://www.theguardian.com/science/2014/mar/04/animal-protein-diets-smoking-meat-eggs-dairy. See also Marsh et al (2012), 'Health Implications of a Vegetarian Diet: A Review".

19   In *Putting Meat on the American Table* (2006) Roger Horowitz describes how fearful animals can affect 'meat value'. He writes: "The methods of an animal's death dramatically affect its flesh... The adrenaline of an animal fearing death creates insufficient lactose acid in the flesh and leads to 'dark cut' meat, an inferior product that does not cure or keep well. It has been a maxim among meat purveyors that livestock should be calm and unaware of their coming demise so that the meat is not damaged by the 'excitement' of a frightened animal" (3).

20   However these figures may be conservative; according to other sources the typical American will eat 21, 000 animals in a lifetime (Stone 2010).

humans' right to dominate nature, including the bodies of animals and their reproductive lives (Luke 2007, Adams 2010, Joy 2010).[21]

Psychologist Melanie Joy uses the term 'carnism' to refer to the "invisible belief system" (or ideology) that propagates meat consumption as "a given, the 'natural' thing to do, the way things have always been and the way things will always be" (29). She claims that "we eat animals without thinking about what we are doing and why because the belief system that underlies this behavior is invisible" (ibid). Vegetarianism is also an ideology but, because it is counter to the dominant meat culture, vegetarianism is visible and noticeable. When an ideology is considered a universal truth, part of 'mainstream' lives, the 'normal' or 'orthodox' way to view things and to be, then it is more likely to become naturalized and accepted as better than all other ways; when it becomes entrenched it also becomes invisible. Joy (2010, 33) argues that:

> While it is difficult, if not impossible, to question an ideology that we don't even know exists, it's even more difficult when that ideology actively works to keep itself hidden. This is the case with ideologies such as carnism. [This] particular type of ideology [is] a violent ideology, because it is literally organized around physical violence. [If] we were to remove the violence from the system—to stop killing animals—the system would cease to exist. Meat cannot be procured without slaughter.

If carnism is the ideology, then 'meat culture' is all the tangible and practical forms through which the ideology is expressed and lived. Meat culture therefore encompasses the representations and discourses, practices and behaviours, diets and tastes that generate shared beliefs about, perspectives on,

---

21   Another way to trace changing dietary patterns and levels of meat consumption across the globe is by analyzing trophic levels. These are synthetic metrics used in ecology to measure a species' position in the food chain. Species' trophic levels are ranked from 1 to 5, with 1 representing primary producers (such as plants and phytoplankton) and 5 representing so-called "apex predators" (polar bears and killer whales) (Bonhommeau 2013, 20617). Using the Food and Agricultural Organization's national data on the human food supply per food item per capita per year (from 1961–2009), fisheries scientist Sylvain Bonhommeau and colleagues recently determined the current Human Trophic Level (HTL) as 2.21, which places humans at the same position as pigs and anchoveta in the food web, and, as the researchers note, "challenges the perception of humans as top predators" (20619). As Bonhommeau et al point out, "[Humans] are closer to herbivore than carnivore", a fact that undermines the naturalization of meat consumption (ibid). However there clearly remains a deep investment in the idea that we are primarily meat eaters, and in the privileging of meat consumption over plant based diets.

and experiences of meat. Like any culture, meat culture is not one thing, nor is it static; it varies widely across and within geographical and cultural locations, as the chapters in this volume will show. While there is a shared general meat culture across industrialized nations—one which maintains the invisible belief system that meat is normal, natural, necessary and nice (known as the 4Ns) (Piazza et al 2015)—different countries, and even places within the same country, will have their own forms of meat culture reflecting regional and social differences such as the ways in which nonhuman species (especially those categorized as killable and edible) are understood and treated.

Contemporary meat culture—the focus of this volume—is also very different from pre-industrial and pre-globalization meat culture, although historical antecedents may affect how particular societies continue to view animals and meat. In newer Western nations the meanings of meat and meat-eating practices may be different from those in older traditional nations. For instance, in 'The American Story of Meat' Barbara Willard (2006) discusses how discourses relating to meat in the context of the United States are closely linked to traditional American values and concepts such as 'rugged individualism', manifest destiny, the biblical story of Genesis (in which humans are granted dominion over the earth), and the notion of meat as a masculine food. "For their efforts at settling the west", Willard (2006, 108) says, "the country rewarded frontiersmen with the myth of the cowboy, a lone hero who took on Indians, wild animals, and the desert heat to build a land suitable for producing what would become America's meal mainstay: beef". Thus, American meat culture is fundamentally associated with (and invested in) capitalism, consumerism and the notion of free will, a perspective that "positions all non-human life as a potential resource" (116).

Meat culture is widespread and ingrained, but it is not without challenge in industrialized nations: the number of people removing meat from their diets or refusing animal products altogether is on the rise. One in eight British adults has now given up eating meat and fish, according to new research by analysts Mintel. Approximately 12% now follow vegetarian or vegan diets, a figure rising to 20% in those aged 16–24. Millions more are so-called 'flexitarians', consciously cutting back on how much meat they consume.[22] This has led to a

---

22   The Meat Free Monday movement, launched by Paul, Mary and Stella McCartney in 2009, encourages people to eat less meat (by committing to one meatless day per week) in order to help slow climate change, protect the environment, reduce the suffering of animals, and improve human health (see http://www.meatfreemondays.com/about/).

thriving market for meat-free products (at £625million in 2013).²³ In the USA, a recent Gallup survey found that close to 5% of Americans refuse meat, with around 8.5 million identifying as vegetarian and 7.5 million as vegan (the number of vegans has doubled since a previous study in 2009).²⁴ The same survey found that omnivores were also reducing their consumption of animal products, with 33% (over 100 million Americans) stating they were choosing to eat meat-free meals more often.²⁵ These trends are backed by Google data showing more and more people are researching vegan diets on-line (Sareem 2013).

Raised awareness of, curiosity about, and commitment to vegetarianism or veganism is no doubt facilitated by the increasing presence in popular culture of realistic and sympathetic veg*n characters in television shows and movies, and also by the publicity generated by celebrity meat refusers such as Bill Clinton, James Cameron, Woody Harrelson, Emily Deschanel, Ellen Page and Peter Dinklage. During an appearance on *Saturday Night Live* in 2015 musician Justin Timberlake 'advertised' veganism to an audience of millions when he sang "Bring it on down to Veganville". Veg*n protagonists and narratives focusing on the paradoxes and cruelties of flesh production and consumption are appearing more in written fiction too (see chapter eight for a close analysis of Michel Faber's millennial novel *Under the Skin*);²⁶ and there has also been a recent flurry of new vegan cookbooks, a rise in the number of veg*n restaurants in Western urban centres, and of veg*n food in supermarkets (in 2014 the German supermarket chain 'Veganz—We Love Life' opened its first branch in the United Kingdom, offering more than 6, 000 vegan products) (Molloy 2013).

Critical Animal Studies (CAS), the more politicized division of Human-Animal Studies, provides a domain of scholarship in which meat culture is explicitly challenged and impeded. Those working in CAS interrogate (with the aim of dismantling) unjust and cruel power relations that are active in humans' relationships with other species (see Wadiwel 2015). Importantly,

23 See http://www.mintel.com/press-centre/food-and-drink/number-of-global-vegetarian-food-and-drink-product-launches-doubles-between-2009-and-2013 Accessed 15 June 2015.
24 A recent poll in New Zealand shows a substantial rise in the number of vegetarians over a 13 year period—from 1–2% of the population in 2002 (Potts and White 2008) to 10.3% in 2016 (http://www.roymorgan.com/findings/6663-vegetarians-on-the-rise-in-new-zealand-june-2015-201602080028). This is despite the nationalist investment in meat and dairy in this country. However NZ veg*ns continue to feel like 'outsiders', due to public admonishment for not supporting the nation's primary industries (Potts et al., 2013).
25 See http://www.gallup.com/poll/156215/consider-themselves-vegetarians.aspx. Accessed 16 June 2015.
26 Other titles include Tobias' *Rage and Reason* (1998), Fitzgerald's *Pigtopia* (2005), D'Lacey's *Meat* (2008), LePan's *Animals* (2009), Lamont's *The Chain* (2013).

CAS scholars also analyze how the domination of nonhuman animals intersects with the marginalization and subjugation of certain humans, thus providing a more thorough and holistic approach to the study of oppression. For example, in her ground-breaking books *The Sexual Politics of Meat* (1990, 2010, 2015) and *The Pornography of Meat* (2003) Carol J. Adams, a trailblazing critic of meat culture, has examined what she calls the '*texts of meat*' in order to reveal the various ways in which "the production of meat's meaning [occurs] within a political-cultural context" (1999, 14). Adams shows how meat culture shapes and is shaped by heterosexist and misogynist rhetoric and images, and also how it relies on the intersecting denigration of women, people of colour, LGBTQI people and nonhuman animals. Chapter two in this volume involves a compelling conversation between Carol Adams and philosopher Matthew Calarco (author of *Thinking Through Animals: Identity, Difference and Indistinction*, 2015) about the intersections and differences between Adams' feminist-vegan critical theory, as exemplified in *The Sexual Politics of Meat*, and French poststructuralist philosopher Jacques Derrida's notion of 'carnophallogocentrism' (introduced in 1991 in '"Eating Well", or the Calculation of the Subject'). In 'Derrida and *The Sexual Politics of Meat*', Adams and Calarco discuss such subjects as carnophallogocentrism and subjectivity, the power of virility and meat eating, and women and animals as overlapping 'absent referents' in meat culture.

Chapter three continues a focus on the absent referent of meat. In 'Rotten to the Bone: Discourses of Contamination and Purity in the European Horsemeat Scandal', sociologists Nik Taylor and Jordan McKenzie analyze around 100 popular media reports on this issue. They argue that while commentators at the time suggested the 'scandal' was a consequence of the food industry's supply chains in a globalized world, or else argued that neo-liberal, free market profiteers were to blame, any critical discussion of *meat eating itself* was strikingly absent from media coverage. Taylor and McKenzie contend that this absence worked to reinforce the normality of meat—and that, in lieu of any critical engagement with meat eating as a practice, discourses of contamination, xenophobia and nationhood dominated coverage of the topic. Employing anthropologist Mary Douglas's theory of purity and danger, the authors also discuss how the European horsemeat scandal demonstrates the significance in meat culture of maintaining hegemonic species boundaries.

Jacqueline Dalziell and Dinesh Wadiwel also examine the connections between meat culture, xenophobia and nationhood in chapter four, 'Live Exports, Animal Advocacy, Race and "Animal Nationalism"'. Australia's lucrative live export trade is under the spotlight here—in particular, the ways in which a recent exposé (by activist group *Animals Australia*) of the abuse to cows shipped from Australia to Indonesian slaughterhouses, resulted in prolonged media coverage and intense public protest. Dalziell and Wadiwel argue powerfully that these energized responses to witnessing the abuse of Australian cattle in another cultural context are saturated by racialized discourse; and that by turning their attention to such atrocities perpetrated in non-Western nations, Australians can ignore or avoid facing the cruelty inherent in their own meat industry at home. Importantly, the authors question whether the outrage about live animal exports evidenced in Australia is really all about the animals, or more about the geopolitics of whiteness.

The next two chapters in this volume are interested in the ways in which meat and dairy industry advertising and marketing shape (and are shaped by) meat culture. In 'The Whopper Virgins: Hamburgers, Gender and Xenophobia in Burger King's Advertising', Vasile Stănescu analyzes a global fast food company's campaign which purported to undertake 'the purest taste test in the world' by transporting hamburgers to remote parts of Thailand, Greenland and Romania, regions where Burger King claimed that people had never encountered such food before, let alone eaten it. BK's promotion received multiple awards and generated significant media interest, culminating in the largest stock price increase in the company's history. In Stănescu's chapter we again see how Western meat culture is imbued with xenophobia. This time, however, there is a focus on American food imperialism, with Stănescu arguing that *The Whopper Virgins* advertisements not only exoticized non-English speaking populations in order to recreate xenophobic reasons for consuming Western style meat, but they also worked to justify Burger King's imposition in these locations of its fast food restaurants as a version of (pseudo-)humanitarianism.

Still on the theme of advertising, chapter six interrogates the Swedish dairy industry's concealment of the cruelty and death inherent in milk production via its springtime family-oriented staged events called 'pasture releases', during which cows (who have been kept inside for autumn and winter months) are returned to paddocks. In 'With Care for Cows and a Love for Milk: Affect and Performance in Swedish Dairy Industry Marketing Strategies', Tobias Linné and Helena Pederson draw upon personal experiences of attending these events in their critique of the ways in which the production and consumption of dairy is strategically hidden from the public eye. They assert such marketing schemes are spectacular products of 'zooesis'—performances masquerading as education, while generating human and animal subjectivities that fit seamlessly within the 'happy meat' and 'new carnivore' regimes of meat culture.

The seventh chapter in this volume redresses the façade of 'happy cows' promoted by dairy industries worldwide. It enables the reader to engage on a personal level with individual cows and steers living on Melissa Boyde's bovine sanctuary outside of Wollongong in Australia. The images and stories of Boyde's much loved companion herd (which includes a 24 year old cow) are juxtaposed with the pictures and histories of industry-owned cattle farmed in the millions for their meat and milk in the Australian context. In '"Peace and Quiet and Open Air": *The Old Cow Project*', Boyde discusses the possibility of photographs amending the unacknowledged brutalities of agribusiness by drawing attention to 'those whose lives are culturally overlooked or deemed unimportant'. With reference to the remains of an abandoned slaughterhouse not far from the author's home (the site still containing rose gardens once designed to provide a peaceful space for the State Abattoir at Homebush's administrators), Boyde also discusses how forgotten or disregarded photographs and objects provide traces of these bloody histories, ensuring we remember, and point to human violence against animals in the present.

Kirsty Dunn's chapter also examines the ways in which stories can expose the real-life suffering of animals in farming systems. In 'Do You Know Where the Light Is?" Factory Farming and Industrial Slaughter in Michel Faber's *Under the Skin*', Dunn analyzes Faber's 2000 novel about a female extraterrestrial called Isserley, whose body has been modified to appear more humanlike for the purpose of picking up male hitchhikers on country roads in Scotland. These men are drugged and then driven to a nearby farm, where they are 'grown out' and 'processed' for consumption by the elite on Isserley's home planet. Through her in-depth examination of representations of intensive farming and slaughter in Faber's novel, Dunn shows how science fiction narratives like *Under the Skin*, which reverse the usual fortunes of animals and humans, can subvert and provoke normative meat culture.

In 'Down on the Farm: Why do Artists Avoid "Farm" Animals as Subject Matter?' artist, animal activist and academic Yvette Watt also examines the representation of farmed animals in cultural forms, this time in art. She reports on a survey she conducted after observing that farm animals were underrepresented in contemporary art, despite the resurgence over the last few decades of artists' interest in animals as serious subject matter for their work. When animals farmed for meat do feature in art Watt noticed that their representation frequently involved their deaths. Focusing on both her own art and that of other artists in whose work farm animals have appeared, this chapter interrogates the connection between the visual representation of 'farm animals' and society's attitude toward them. Watt proposes that farm animals, whose form is continuously changed by human intervention, bridge the nature-culture dichotomy, as they are seen simultaneously as natural *and* 'human-made', with this confusion of category resulting in 'farm' animals being seen as less worthy subjects for artists. She argues that the lack of respectful representations of farm animals in the visual arts reflects the pervasiveness of a meat culture that has reduced 'farm' animals to the status of commodity, further diminishing their inherent worth, and reflecting the fact that artists are as much a part of 'meat culture' as society in general.

In chapter ten Karen Davis, president of the world's foremost advocacy organization for exploited birds, United Poultry Concerns, critiques the practice of creating public empathy and support for the 'rights' of birds and other animals by comparing their levels of understanding and awareness to those of children or of intellectually compromised humans (this has been one approach in Peter Singer's work, for example). Drawing upon her own experiences of caring for birds over twenty years, Davis shows in 'The Provocative Elitism of "Personhood" for Nonhuman Creatures in Animal Advocacy Parlance and Polemics' that chickens, in particular, are conscious, emotional and intelligent beings with adaptable sociability and a range of intentions and personalities. With reference to insights provided by cognitive ethology, she relates personal stories about the birds at her avian sanctuary in order to demonstrate how erroneously and unfairly these beings are represented and treated in meat culture.

Chapters eleven, twelve and thirteen have in common an engagement with veganism, which might be described as carnism's 'counter culture'. Matthew Cole and Kate Stewart begin with 'Why Isn't the Doctor Vegan? The Irruption and Suppression of Vegan Ethics in *Doctor Who*', which explores the tensions between challenging and reproducing the exploitation of other animals as food in the internationally popular BBC (British Broadcasting Corporation) television series, *Doctor Who*, which celebrated its 50th anniversary in 2013. The analysis in the chapter, which is informed by the authors' own conceptual model that 'maps' Western cultural legitimation of the exploitation of other animals (Cole and Stewart 2014), focuses on the ways in which this programme suppresses a logical tendency towards vegan ethics. Despite the brief appearance of a vegetarian Doctor in episodes in the 1980s, Cole and Stewart claim that the on-screen meat-eating practices of Doctors since sit uncomfortably with the moral logic of this character's consistent opposition to domination and exploitation on the basis of claims to 'superiority'. While the newest series of *Doctor Who* exemplifies the mainstream cultural suppression of discomfiting ethical challenges to conventional exploitative food practices, Cole and Stewart argue that it nevertheless retains the potential to be a platform for a character-driven exploration and implicit advocacy of vegan ethics in popular culture.

For vegetarians or vegans living with carnivorous or omnivorous companion animals the question of what to feed these dependents is a constant dilemma. Sociologist Erika Cudworth's chapter, 'On Ambivalence and Resistance:

Carnism and Diet in Multi-species Households', draws upon an empirical study of human-canine relationships in her examination of the pet-food industry, which is of course fundamentally connected to the global meat and dairy industries. She argues that carnist ideology pervades discourses of care and concern for much loved dogs, and that ambiguous relations to the animals such dogs eat are also often expressed. Cudworth also shows how, even for vegetarian and vegan 'dog owners', and with the availability of varieties of meat-free pet foods now, the influential discourses associated with meat culture can reinscribe meat eating practices for the nonhuman members of multi-species households.

Chapter thirteen, 'Vegan Transition and the Relationship Context', also involves empirical research. During 2013, in the context of a larger project on sustainable veganism, sociologist Richard Twine interviewed forty vegans across three cities in the United Kingdom in order to explore how, in the context of dominant meat culture, vegan practice raises tensions within familial relationships and friendships (and how such conflicts are negotiated). This chapter is informed by practice theory, which, Twine explains, focuses on how practices (rather than individuals) consolidate and change with respect to three key elements: competency (skills and know-how), materials (the body and objects or technologies that make up a practice) and meaning (ideas, symbolic meanings and norms). Twine's analysis of interview material offers fascinating insights into how the 'normalization' of vegan practice is fostered and achieved within the context of close relationships with non-vegans.

In chapter fourteen we are introduced to the emerging field of Critical Plant Studies. At first glance we might assume common ground between this new area and Critical Animal Studies, each domain seemingly focused on disturbing anthropocentric forms of knowledge and the privileging of human experience. However, as Greta Gaard explains in her timely and powerful essay, 'Vegetal Ecocriticism and the "Critical Plant Studies" Backlash', vegan feminist ecocriticism (which questions ecocritics who continue to consume meat and other animal products) has recently encountered a backlash in the form of *'vegetal ecocriticism'*—a component of Critical Plant Studies. Gaard asserts that by claiming there is a political line between plants and animals—akin to the human/animal divide—and that plant-life and animals share similar modes of sentience, vegetal ecocriticism actually seeks to delegitimize the very real suffering of animal species, and place human food choices on a terrain of moral relativism. Importantly, Gaard's chapter provides cogent responses and strategies that vegan feminist ecocritics can employ when countering such claims from within vegetal ecocritism.

The fourteen chapters comprising this volume thus provide multiple detailed analyses from across a range of academic affiliations and geographical locations, each author concerned to intercept and disrupt the taken-for-grantedness, pervasiveness and callousness of the animal industrial complex, carnism, and contemporary meat culture.

References

Adams, Carol J. *The Sexual Politics of Meat: A Feminist-Vegetarian Critical Theory.* New York: Continuum, 1990, 2010, 2015.
Adams, Carol J. *The Pornography of Meat.* New York: Continuum, 2003.
Adams, Carol J. "Why Feminist-Vegan Now?" *Feminism & Psychology*, 20:3 (2010): 302–317.
Alexander, James. "The Unlikely Origin of Fish and Chips". *BBC News*, 2009. http://news.bbc.co.uk/2/hi/8419026.stm, accessed 15 June 2015.
Balcombe, Jonathan. *What a Fish Knows: The Inner Lives of Our Underwater Cousins.* New York. Farrar, Straus & Giroux, 2016.
Bonhommeau, Sylvain et al. "Eating Up the World's Food Web and the Human Trophic Level". PNAS, 110:51 (2013): 20617–20620.
Braithwaite, Victoria. *Do Fish Feel Pain?* Oxford: Oxford University Press, 2010.
Cole, Matthew. "From 'Animal Machines' to 'Happy Meat'? Foucault's Ideas of Disciplinary and Pastoral Power Applied to 'Animal-Centred Welfare Discourse'". *Animals*, 1 (2011): 83–101.
Cole, Matthew and Stewart, Kate. *Our Children and Other Animals: The Cultural Construction of Human-Animal Interaction in Childhood.* Farnham: Ashgate, 2014.
Cook, Christopher D. "Sliced and Diced: The Labor you Eat". In *The CAFO Reader: The Tragedy of Industrial Animal Factories*, edited by Daniel Imhoff. Los Angeles: University of California Press, 2010: 232–239.
D'Lacey, Joseph. *Meat.* Fyfield, Oxfordshire: Oak Tree Press, 2008.
Derrida, Jacques. *The Animal That Therefore I Am.* Edited by Marie-Louise Mallet. Translated by David Wills. New York: Fordham University Press, 2008 (2006).
Eisnitz, Gail. *Slaughterhouse: The Shocking Story of Greed, Neglect and Inhumane Treatment Inside the US Meat Industry.* New York: Prometheus Books, 2007.
Fitzgerald, Amy J. and Taylor, Nik. "The Cultural Hegemony of Meat and the Animal Industrial Complex". In *The Rise of Critical Animal Studies: From the Margins to the Centre*, edited by Nik Taylor and Richard Twine. London: Routledge, 2014: 165–182.
Fitzgerald, Kitty. *Pigtopia.* London: Faber, 2005.
Foer, Jonathan Safran. *Eating Animals.* Penguin UK, 2010.
Fonterra. "The Global Dairy Industry 2015". www.fonterra.com/nz/en/Financial/Global+Dairy+Industry, accessed 29 May 2015.
Food and Agriculture Organization of the United Nations. *Livestock's Long Shadow: Environmental Issues and Options.* FAO, 2006. ftp://ftp.fao.org/docrep/fao/010/a0701e/a0701e.pdf, accessed 29 May 2015.

Food and Agriculture Organization of the United Nations Fisheries and Aquaculture Department. *Global Statistics*. FAO, 2015. http://www.fao.org/fishery/statistics/en, accessed 29 May 2015.

Fulton, Tim. "Filipino Dairy Rights Group Fighting Back". *Farmers Weekly News*. http://farmersweekly.co.nz/article/filipino-dairy-rights-group-fighting-back, accessed 15 January 2016.

Galobardes, Bruna, Morabia, Alfredo, and Bernstein, Martine S. "Diet and Socioeconomic Position: Does the Use of Different Indicators Matter?" *International Journal of Epidemiology*, 30:2 (2001): 334–340.

Greger, Michael. In *Mad City Chickens*, directed by Tashai Lovington and Robert Lughai. Tarazod, 2009.

Horowitz, Roger. *Putting Meat on the American Table: Taste, Technology, Transformation*. Baltimore: The Johns Hopkins University Press, 2006.

Hovorka, Alice J. "Transspecies Urban Theory: Chickens in an African City". *Cultural Geographies*, 15 (2008): 119–141.

Hovorka, Alice J. "Women/Chickens v. Men/Cattle: Insights on Gender-Species Intersectionality". *Geoforum*, 43:4 (2012): 875–884.

Heinrich Boll Foundation and Friends of the Earth Europe. *Meat Atlas: Facts and Figures About the Animals We Eat*. https://www.boell.de/en/meat-atlas, accessed 29 May 2015.

Hoag, Hannah. "Humans are Becoming More Carnivorous". *Nature News*, December 2 2013. http://www.nature.com/news/humans-are-becoming-more-carnivorous-1.14282, accessed 16 June 2015.

Humane Society of the United States. "Slaughter Totals". http://www.humanesociety.org/news/resources/research/stats_slaughter_totals.html, accessed 3 June 2015.

Joy, Melanie. *Why We Love Dogs, Eat Pigs and Wear Cows: An Introduction to Carnism*. San Francisco: Conari, 2010.

Kimbrell, Andrew. "Cold Evil: The Ideologies of Industrialization". In *The CAFO Reader: The Tragedy of Industrial Animal Factories*, edited by Daniel Imhoff. Los Angeles: University of California Press, 2010: 29–43.

Lamont, Robin. *The Chain*. USA: Grayling Press, 2013.

Lappe, Anna. "Diet for a Hot Planet: Livestock and Climate Change". In *The CAFO Reader: The Tragedy of Industrial Animal Factories*, edited by Daniel Imhoff. Los Angeles: University of California Press, 2010: 240–250.

LePan, Don. *Animals: A Novel*. Berkeley, CA: Soft Skull Press, 2010.

Liying, Ahang et al. "Present and Future of China's Broiler Industry". In *Poultry Beyond 2005*, edited by R. J. Diprose et al. Christchurch, 2001: 21–31.

Luke, Brian. *Brutal: Manhood and the Exploitation of Animals*. Chicago: University of Illinois Press, 2007.

Marcus, Erik. *Meat Market: Animals, Ethics, and Money*. Boston: Brio Press, 2005.

Marsh, Kate A., Zeuschner, Carol L. and Saunders, Angela V. "Health Implications of a Vegetarian Diet: A Review". *American Journal of Lifestyle Medicine*, 6:3 (2012): 250–267.

Meadows, Richard and Cronshaw, Tim. "Wages Breaches Found in Dairy Farm Investigations". *NZ Farmer*, 1 April 2015. http://www.stuff.co.nz/business/farming/dairy/67577313/Wages-breaches-found-in-dairy-farm-investigations, accessed 15 January 2016.

Mendelson, Anne. "The Milk of Human Unkindness: Industrialization and the Supercow". In *The CAFO Reader: The Tragedy of Industrial Animal Factories*, edited by Daniel Imhoff. Los Angeles: University of California Press, 2010: 131–138.

Mizelle, Brett. *Pig*. London: Reaktion, 2011.

Molloy, Antonia. "No Meat, No Dairy, No Problem: Is 2014 the Year Vegans Become Mainstream?" *The Independent*, Saturday 13 December 2013. http://www.independent.co.uk/life-style/food-and-drink/features/no-meat-no-dairy-no-problem-is-2014-the-year-vegans-become-mainstream-9032064.html, accessed on 15 June 2015.

Mood, A. and Brooke, P. "Estimating the Number of Farmed Fish Killed in Global Aquaculture Each Year". http://fishcount.org.uk/published/std/fishcountstudy2.pdf (2010), accessed 13 June 2015.

Noske, Barbara. *Humans and Other Animals: Beyond the Boundaries*. London, England: Pluto Press, 1989.

Pachirat, Timothy. *Every Twelve Seconds: Industrialized Slaughter and the Politics of Sight*. New Haven: Yale University Press, 2011.

Packwood Freeman, Carol. "This Little Piggy Went to Press: The American News Media's Construction of Animals in Agriculture". *The Communication Review*, 12.1 (2009): 78–103.

Pan, A. et al. "Red Meat Consumption and Mortality: Results from Two Prospective Cohort Studies". *Archives of Internal Medicine*, 172:7 (2011): 555–563.

Parry, Jovian. *The New Visibility of Slaughter in Popular Gastronomy*. Unpublished MA Thesis (Cultural Studies), University of Canterbury, 2010.

Piazza, J. et al. "Rationalizing Meat Consumption: The 4Ns". *Appetite*, 91 (2015): 114–28.

Pilgrim, Karyn. "'Happy Cows', 'Happy Beef': A Critique of the Rationales for Ethical Meat". *Environmental Humanities*, 3 (2013): 111–27.

Pleasance, Chris. "We Really Are a Nation of Meat Eaters: Carnivores Devour More Than 7,000 Animals In Their Lifetime Including 11 Cows, 2,400 Chickens and 30 Sheep". *The Daily Mail*, 9 March 2015. http://www.dailymail.co.uk/news/article-2985910/We-really-nation-meat-eaters-Carnivores-devour-7-000-animals-life-time-including-11-cows-2-400-chickens-30-sheep.html, accessed 1 June 2015.

Potts, Annie. *Chicken*. London: Reaktion, 2012.

Potts, Annie and Parry, Jovian. "Vegan sexuality: Challenging heteronormative masculinity through meat-free sex". *Feminism & Psychology*, 20.1 (2010): 53–72.

Potts, Annie and White, Mandala. "New Zealand Vegetarians: At Odds With Their Nation". *Society & Animals*, 16.4 (2008): 336–353.

Potts, Annie, Armstrong, Philip and Brown, Deidre. *A New Zealand Book of Beasts: Animals in our Culture, History and Everyday Life*. Auckland: Auckland University Press, 2013.

Robb, D. H. F. and Kestin, S. C. "Methods Use to Kill Fish: Field Observations and Literature Reviewed". *Animal Welfare*, 11 (2002): 269–282.

Rogers, Lesley J. *The Development of Brain and Behaviour in the Chicken*. Wallington, CABI, 1995.

Ruby, Matthew B. "Vegetarianism. A Blossoming Field of Study". *Appetite* 58.1 (2012): 141–150.

Sareem, Anjali. "Interest in Vegan Diets on the Rise: Google Trends Notes Public's Increased Curiosity in Veganism". *The Huffington Post*, 3 March 2013. http://www.huffingtonpost.com/2013/04/02/interest-in-vegan-diets-on-the-rise_n_3003221.html, accessed 4 June 2015.

Serpell, James. *In the Company of Animals: A Study of Human-Animal Relations*. Cambridge: Cambridge University Press, 1996.

Stone, Gene. (Ed). *Forks Over Knives: The Plant-based Way to Health*. New York: The Experiment, 2011.

Taylor, Nik. *Reversing Meat-eating Culture to Combat Climate Change*. Haslemere: World Preservation Foundation, 2012.

Taylor, Nik. "Suffering is Not Enough: Media Depictions of Violence to Other Animals and Social Change". In *Critical Animal and Media Studies: Communication for Nonhuman Animal Advocacy*, edited by Nuria Almiron, Matthew Cole and Carrie Packwood Freeman. New York: Routledge, 2016: 42–55.

Tobias, Michael. *Rage and Reason*. Oakland, CA: AK Press, 1998.

Twine, Richard. *Animals as Biotechnology: Ethics, Sustainability and Critical Animal Studies*. London: Earthscan, 2010.

United States Environment Protection Agency—Office of Research and Development. "Risk Assessment Evaluation for Concentrated Animal Feeding Operations." (2004). http://nepis.epa.gov/Exe/ZyPURL.cgi?Dockey=901V0100.txt, Accessed on 20 May 2015.

Vis, V. D. and Kestin, S. C. "Killing of Fishes: Literature Study and Practice-Observations". (Field Research) Report Number C 037/96, RIVO DLO, 1996.

Wadiwel, Dinesh. *The War Against Animals*. Leiden & Boston: Brill Rodopi, 2015.

Willard, Barbara E. "The American Story of Meat." *Journal of Popular Culture*, 6:1 (2006): 105–118.

Young, Rosamund. *The Secret Lives of Cows*. London: Farming Books & Videos Ltd, 2003.

CHAPTER 2

Derrida and *The Sexual Politics of Meat*

Carol J. Adams and Matthew Calarco

What sorts of intersections, associations, or refusals might be found between Jacques Derrida's work and *The Sexual Politics of Meat*? Carol and Matt decided that a discussion format might enable them to explore answers to this question. What follows represents a part of that larger conversation. While Carol's work in decoding the sexual politics of meat stretches over more than thirty years, expanding into close consideration of the functioning of representations, Derrida's elaboration on his neologism *carnophallogocentrism* remained largely suggestive and deferred. Carol and Matt recognize that this difference, among many other reasons, means that their goal never was to compare and contrast but to create a dialogue and exploration. They wish to thank Vasile Stănescu for suggesting that they explore this topic together.

Matt:
One of the more evident points of contact between your work and Derrida's writings on animals can be found in (1) your overarching project of critically examining and contesting the sexual politics of meat, and (2) Derrida's occasional attempts to think through the connections between subjectivity, sexism, and eating meat by way of his concept of *carnophallogocentrism*. In order to explore this overlap between the sexual politics of meat and carnophallogocentrism, it might be useful for me to lay out a few of Derrida's ideas in a bit more detail and suggest some points at which your and his project overlap and diverge.

It is important to note that Derrida's work on carnophallogocentrism (and animals more generally) was, despite his occasional protestations to the contrary, never in the foreground in the same manner that the sexual politics of meat is in your work. That Derrida nearly always deferred attention from questions concerning carnophallogocentrism and animals is indicative of both a certain caution and also (I would suggest) a lack of a sense of urgency in his writings. He always found time to write on other pressing socio-political issues and develop his positions in great detail on many of those issues; but when issues concerning animals and other nonhuman beings arose, he most often held any careful analysis of such matters in abeyance. There can be little doubt that Derrida was cautious when approaching issues surrounding animals and non-human life primarily because of the sheer difficulty and magnitude required for a full treatment of the topic. But his tendency to hold questions about animals in abeyance was perhaps not just a symptom of this caution and hyper-prudence. It is also clear—or at least, nearly everything in his writings and political activity would suggest—that the transformation of the living conditions of many animals as well as the transformation of human relationships with animals simply was not one of his over-arching priorities in the same way it is for your work or mine.

So, allow me to start off with a basic discussion of the concept of carnophallogocentrism and how it fits into Derrida's work. He mentions this concept in several of his writings, but his most sustained examination of it occurs in a 1988 interview with Jean-Luc Nancy entitled "'Eating Well,' or the Calculation of the Subject" (1991, 113–14). The interview is Derrida's contribution to Jean-Luc Nancy's attempt to take stock of recent work in so-called post-humanist thought (that is, thinking that proceeds from the critical interrogation of 'humanism,' or what it means to be a human 'self' or 'subject' in the Western philosophical tradition). Throughout the interview, Derrida repeatedly makes the point that, despite the seemingly radical and thoroughgoing critique of selfhood and subjectivity in recent Continental philosophy (and in Heideggerian and Levinasian thought in particular), insufficient attention has been paid both to the anthropocentric nature of dominant Western philosophical conceptions of subjectivity and also to the lingering anthropocentrism in the more cutting-edge, post-humanist critiques of subjectivity. In other words, he detects a certain dogmatic adherence to anthropocentrism even among his more sophisticated fellow post-humanist critics. It is in this context that Derrida tries to distance himself from dogmatic anthropocentrism by calling attention to the carnophallogocentric constitution of human subjectivity in the Western philosophical and cultural traditions.

Derrida's earliest writings aimed to expose the *logo*-centric assumptions of these traditions (logocentrism here denoting the privileges and priorities granted by Western philosophy to the rational, self-aware, self-present, speaking subject).[1] And, when his attention turned to issues dealing more directly with sexuality and gender, he tried to demonstrate the inextricable linkages between logo-centrism and *phallo*-centrism (phallocentrism here denoting the quintessentially virile and masculine aspects of Western social institutions and conceptions of subjectivity), leading him to use the neologism

---

1 See especially Part I of Jacques Derrida, *Of Grammatology* (Baltimore, MD: Johns Hopkins University Press, 1976).

phallogocentrism to denote these joint phenomena.[2] In "'Eating Well,'" Derrida suggests that *carno* should be added to *phallogocentrism* in order to emphasize that the notion of the subject that is being critiqued in post-humanist thought should be understood not simply as a fully self-present, speaking, masculine subject but also as a quintessentially *human, animal-flesh-eating* subject.

By the late 1980s, then, Derrida is arguing that the critical deconstruction of subjectivity should be seen as a critical deconstruction of carnophallogocentrism. This project calls for an intersectional analysis of at least three coordinates or registers in the constitution of subjectivity:

- self-presence (the *logos* of self-mastery, reason, speech, and transparent, unmediated access to one's inner mental life);
- masculinity (the manner in which virile and masculine ideals are infused throughout and dominate the socio-cultural order); and
- carnivorism (the requirement of the literal and symbolic consumption of flesh, a commitment to anthropocentrism, the hierarchical ranking of human subjects over non-human animals)

While there are significant differences between the sexual politics of meat and the deconstructive analysis of carnophallogocentrism as I have initially explained it here, I wonder if we might first turn to a discussion of how your project has certain *positive* affinities with Derrida's work.

Carol:
I have a sense that in its analysis, *The Sexual Politics of Meat* intersects with 'carnophallogocentrism' in several ways. Derrida was attempting to name the primary social, linguistic, and material practices that go into becoming a subject within the West and how explicit carnivorism lies at the heart of classical notions of subjectivity, especially male subjectivity. Similarly, I have been trying to show how a feminist analysis that decenters male subjectivity and challenges a violence long associated with human male behavior is impelled to include a critique of carnivorism, too. I argue that a challenge to the male-defined Western subject needs to include challenging the foods that are assumed to be 'his' foods. In this, I make clear that I am not talking just about 'men' but how everyone in the West is implicated by the sexual politics of meat.

---

2 The "indissociability" of logocentrism and phallocentrism are discussed most lucidly by Derrida in "'This Strange Institution Called Literature': An Interview with Jacques Derrida," in Derek Attridge, ed., *Acts of Literature* (London: Routledge, 1992), 57–60.

I want to suggest that the carnophallogocentric subject is the subject created by a culture with the foundational premise of the sexual politics of meat. In other words, the sexual politics of meat is constituting this carnophallogocentric subject at many levels. I am not claiming this is the only force at work; I am asserting its influence is not negligible and needs to be recognized.

Matt, you identify three coordinates or registers in the construction of subjectivity that constitute, in a sense, the carnophallogocentric subject. Similarly, I propose several aspects to *The Sexual Politics of Meat*. It is not one 'thing,' one quality, one 'fact,' it is, rather, kaleidoscopic and shifting in how we experience it, but at the minimum contains these parts:

- The association of virility and meat eating
- The functioning of the structure of the absent referent
- Women and animals positioned as overlapping absent referents in a patriarchal culture

Virility and Meat Eating

In *The Sexual Politics of Meat*, I argue that a link exists between meat eating and notions of masculinity and virility in the Western world. Meat eating societies gain human male identification by their choice of food, creating and recreating an experience of male bonding in various male-identified locations, such as steak houses, fraternities, strip clubs, or (domesticated) at a barbecue.

Meat eating bestows an idea of masculinity on the individual consumer. Popular culture manifestations of the sexual politics of meat can be found imbricated throughout various media and in personal behavior. Generally, they imply that a man needs meat and that a woman should feed him meat. From French commercials to newspaper advertisements for Father's Day, the theme is reiterated.

Meat eating is an act of self-definition as a privileged (male-identified) human.

A belief exists that strength (male-identified) comes from eating 'strong animals' (for instance, 'beef'), and that vegetables represent passivity. Thus, conventionally, vegetarianism was considered appropriate for women and anyone associated with women. These ideas, which appear in the first chapter of *The Sexual Politics of Meat* seem to resonate with Derrida's idea of "carnivorous virility." A bumper sticker like "Eat Beef. The West Wasn't Won on Salad" exemplifies this attitude. In one statement, it is putting down foods associated with women, elevating animal foods, and at its heart, celebrating the genocide of Native Americans.

The issue is not only human exceptionalism in the myriad ways it is recuperated to justify eating animals; it is how the human is conceived, as male-identified, with a male-identified diet. A 2006 Hummer advertisement features a man buying tofu in a supermarket. Next to him a man is buying gobs of raw meat. The tofu-buying man notices this and becomes alert to and anxious about his virility, apparently compromised by his tofu-buying. He hurries from the grocery store and heads straight to a Hummer dealership. He buys a new Hummer and is shown happily driving away, munching on carrot. The original tag line for the ad was "Restore your manhood" (Stevenson 2006). (It was changed to "restore the balance"). The implication that the Hummer acts as compensation for the vegetarian man's failure to eat manly protein suggests that one aspect of culture committed to carnophallogocentric subjectivity is the belief in the logic of the sexual politics of meat.

Since *The Sexual Politics of Meat* was published, I have noticed that many popular culture appeals to men (especially white, heterosexual men as in the Hummer ad) seem to be rebuilding what feminism and veganism have threatened. In terms of *The Sexual Politics of Meat*, we see several recuperative responses that seek to reinstate manhood, meat eating, and both interactively. From unsophisticated wall paintings on restaurants to slick Superbowl commercials, the message that meat's meaning is expressed through sexual politics is constantly recreated. The ads that I examine in *The Sexual Politics of Meat Slide Show* appeal to, reassure, flatter, massage, reinforce the carnophallogocentric subject Derrida espied.

The Absent Referent I

Behind the *carno* in carnophallogocentrism is the absent referent. Through butchering, animals become absent referents. Animals in name and body are made absent *as animals* for meat to exist. Animals' lives precede and enable the existence of meat. If animals are alive they cannot be meat. Thus a dead body replaces the live animal. Without animals there would be no meat eating, yet they are absent from the act of eating meat because they have been transformed into food.

The absent referent is that which separates the flesh eater from the animal and the animal from the end product. The function of the absent referent is to allow for the moral abandonment of a being. In many quotidian ways, the absent referent functions to cloak the violence inherent to meat eating, to protect the conscience of the meat eater, and to render the idea of individual animals as immaterial in the face of someone's specific and selfish desires to consume them.

Implicit in Derrida's use of *carno* is sacrifice; but *explicit* in his use is '*carn*ivorism,' in other words, flesh eating. What the concept of the absent referent uncovers is both the fact that animals die individually, and that we wish to keep hidden what we are doing to animals. We cannot lose track of the fact that flesh eating occurs through an act of violence that the 'carnist,' in Melanie Joy's (2010) terms, accomplishes through an activity of consumption.

The act of killing animals (like the act of eating meat) is part of the project of constructing carnophallogocentric subjectivity. Isn't the violence underlying the act an important aspect of both 'sacrifice' in Derrida's term and in associating the act with phallocentrism? Recently, scholars have begun studying the 'New Carnivore' or 'neocarn' movement, in which popular food shows (Gordon Ramsay's *The F Word* and Jamie Oliver's *Fowl Dinners*) display the killing and consumption of domesticated animals the celebrity chefs have raised (Parry 2010, 381–96). The need to reassure masculinity is an unstated project of these televisions shows—a sort of desperate performative rebuilding of the carnophallogocentric subject through violence. They want one sort of honesty (killing) and hide behind a greater dishonesty. The need to make the kill present is a hypermasculine reinscription of the sexual politics of meat.

Let us say "the structure of the absent referent is xyz" (the literal death of the animal, the hiding of the facts of that death, the lifting of the animals' death to a higher meaning through metaphor and consumption). With these television shows, we see x and z still functioning (the objectifying, the eating of a dead object) but y is not absent; it has been made demonstrably present (the death is not hidden). Why 'y'? And the answer turns on the issue of the instantiation of the human male subject who, in these instances, requires not just consumers of goods, but consumers of their actions, in a sense, voyeurs.

The reiterative nature of the sexual politics of meat (finding new ways of reinforcing 'carnivorous virility') suggests how fundamental it is to the operation of Western culture.

Women and Animals as Overlapping Absent Referents

In *The Sexual Politics of Meat*, I argue that women and animals are overlapping absent referents. This is tied to another aspect of the absent referent, when a dominant and domineering language consumes and negates the violent transformation of living to dead and 'lifts' that experience into metaphor. Then it is applied to vulnerable or otherwise disenfranchised beings.

This structure of overlapping absent referents also moves in the other direction as well, in which women's objectification becomes the basis for cultural constructions about meat animals. It was difficult to find a sound bite for this theory, but by the time I wrote *The Pornography of Meat* I had one: in a patriarchal, meat eating world animals are feminized and sexualized; women are animalized.

In terms of overlapping cultural images involving animals and women, things have gotten worse. Meat advertisements that sexualize and feminize animals have been around for more than thirty years, and during this time, they have become more widespread and more explicit. What *Hustler* pornographically imagined women as thirty-five years ago, Burger King, Carls' Jr, and many other dead animal purveyors recreate and suggest now. You can find *Hustler's* image of a woman going through a meat grinder image prettified in an ad for the HBO series *The Comeback* featuring Lisa Kudrow. Burger King takes the *Hustler* mentality—women as meat, as hamburger, and stylizes it for Super Bowl commercials. The 2009 *Sports Illustrated* swimsuit issue ("Bikinis or Nothing") includes an ad for Arby's with hands removing two hamburger buns as though they are taking off a bikini top.

In *The Sexual Politics of Meat*, I say that the connections between women and animals that I am drawing are contingent and historical. But I argue that theoretically speaking, politically speaking, these contingent historical overlaps (in which the animal substitutes for the woman, and the woman, or part of a woman, substitutes for a dead animal) are relevant conjunctions to make.

I propose that in general women are visually consumed; animals literally consumed. Then, I push on just *who* we are consuming. This may be one place where I make explicit something that may be only implicit in Derrida. If the carnophallogocentric subject knows 'himself' to be a subject through the inflection of meat eating and male-centeredness, then implicitly there are objects in 'his' life that contribute to the creation of 'his' subjectivity.

I am interested in those *objects*, and who our culture allows to become those objects, and how the process of objectification is working. In my work, I found an overlap of cultural images of sexual violence against women and fragmentation and dismemberment of nature and the body in Western culture. I propose that a cycle of objectification, fragmentation, and consumption linked butchering with both the representation and reality of sexual violence in Western cultures, that normalizes sexual consumption. I believe this aspect of my work ties in with Derrida's concern about "commonly accredited oppositional limits" (2009, 36). If I can appropriate Derrida's term to explicate my ideas, carnophallogocentric subjectivity is invested in the oppositional framework of Western culture, and benefits from the lowering/debasing of some beings that accompanies this oppositional framework. But what I try to show is how the oppositional points (human/animal, man/woman for instance) are intensified by linking two parts of the negated side—female and animal. I do not believe these points are fixed, nor do I necessarily simply valorize the side that has been lowered. I am concerned with the dynamics of linking that occur on both sides.

This linking is important, because the carnophallogocentric subject is constituted, in part, through the power to objectify living beings, to make other subjects into objects. This subject whose qualities of self-mastery, reason, speech, etc., feel inherent—is not one way that this subject *knows* his (or her) self-mastery, reason, speech etc. precisely through the objectification of other beings?

I remember when O. J. Simpson was first suspected of murdering his wife. Often, the comments that I heard were that he was such a 'charismatic' person. No one seemed to stop and think perhaps there was a link between his *charisma* and his battering behavior. He could behave the way he did in public because of what he was doing, controlling another person, in private. Similarly, is not one aspect of the construction of the carnophallogocentric subject the 'consumption' or 'sacrifice' (forced, imposed, not selected) of other beings to his (or her) needs or wants?

One of the things *The Sexual Politics of Meat* was trying to do was to capture the dynamics of this.

Matt:
Your remarks here on the contingent and historical aspects of the sexual politics of meat serve as an excellent starting point to carry our discussion forward. I would like to use those remarks to press on the question of whether this kind of attention to the sexual politics of meat in your work and carnophallogocentrism in Derrida's work is itself in need of a supplement.

Carnophallogocentrism as *the* Dominant Schema of Subjectivity

When Derrida turns in the late 1980s to a discussion of carnophallogocentrism, this turn is the direct result of his earlier efforts to pinpoint both the dominant tendency of metaphysics and its unthought ground. The dominant tendency of metaphysics for Derrida is to elaborate and privilege a certain thought of presence and identity understood most often as a self-present, self-conscious subject. In this gesture, he is quite close to Heidegger. But where Derrida critically and importantly departs from Heidegger is in his characterization of subjectivity as specifically involving *human, masculine,* and *carnivorous* dimensions. These dimensions were not a point of focus for Heidegger, nor are they for most neo-Heideggerians. That Derrida takes these specific dimensions seriously is very much to his credit, as he is one of the very few Continental philosophers who have noticed not just the phallocentric aspects of metaphysical notions of subjectivity but also the anthropocentric and carnivorous tendencies of that tradition.

But I think Derrida also overplays his hand a bit on this point. Although his remarks on carnophallogocentrism are in no way programmatic, they are nevertheless intended to capture the deep structure, the "dominant schema" of subjectivity of Western metaphysics. And it is precisely here, in the attempt to name definitively the dominant and quintessential form of subjectivity that I think he overreaches. Although I would readily and enthusiastically agree that the concept of carnophallogocentrism brings together important and significant trends in so-called Western metaphysics, I doubt that it captures *the* dominant schema of metaphysical subjectivity. Are there not other important trends and tendencies to add to this hyphenated list? Why limit our analysis to logocentrism, phallocentrism, and carnivroism? Why do these three registers have priority?

The point here is not that this list should be extended to 4+n registers in order to be made accurate and complete. The list of forces and relations that constitute us as subjects can obviously never be made complete. Instead, I am wondering whether we should subscribe to this neo-Heideggerian logic of dominant schemas in the first place. Is there really something called 'Western metaphysics' whose dominant schema can be uncovered? And even if we were to arrive at such knowledge about the inner workings of Western metaphysics, what is the ultimate wager or hope behind this approach? What are the critical and transformative implications for uncovering carnophallogocentrism as the dominant schema of Western metaphysics? We might also ask: How would this kind of neo-Heideggerian approach relate to other critical analyses of Western culture that differ substantially in their focus and strategies (I am thinking, for example, of indigenist or Marxist critical approaches that often have very different premises and critical aims from those of deconstruction).

By setting up things in this neo-Heideggerian way (the search for dominant schemas, definitive limits of metaphysics, and so on), Derrida ends up (whether intentionally or not) framing the philosophical and critical task in such a way as to foreclose other analyses and to cut off linkages with other approaches. By contrast with Derrida, I would much prefer to understand the discussion of carnophallogocentrism and the sexual politics of meat as but one way to uncover a force that is (to use your words) "not negligible and needs to be recognized." If we begin from this space and in this kind of modest theoretical spirit, it creates the possibility for additional linkages between critical analyses and for allowing such analyses to mutually inform one another in a critical and progressive manner.

Ultimately, the problem I have with Derrida's analysis of carnophallogocentrism is that it remains at bottom almost entirely intra-philosophical and intra-theoretical. To explain this point further: What I find important about the concept of carnophallogocentrism is that it is, among other things, useful for (1) capturing important tendencies in our culture surrounding the constitution of 'properly' human subjects, and (2) suggesting possible linkages among various critical perspectives and movements for social transformation (in this case, feminism and movements on behalf of animals). But the latter considerations are, at best, secondary for Derrida and always deferred in favor of his primary interest. His primary interest seems to lie in a careful philosophical analysis of the deep structures of metaphysics (we find hundreds upon hundreds of pages dedicated to careful readings of Heidegger, Levinas, Lacan, Descartes, Kant, and so on)—and not in creating alliances across radical and critical practices.

This point should not be taken to suggest that Derrida is entirely uninterested in radical politics and social transformation; but it is hard to resist the conclusion that such interests remain in the background. Any bridge that might be built between his preferred method of patient philosophical analysis and radical political transformation is almost always deferred in his work, much like 'the question of the animal' is often deferred. And when he is pressed on the connections between deconstruction and social transformation, he most often dodges the question by implying that such questions are driven by those who expect a political 'program' from him. But it is not the case that I or others expect or desire a program from him (I doubt anyone who is involved in this discussion or reads this material finds political programs desirable); instead, the question that is at issue here is how this kind of critical work informs or is informed by alternative modes of practice and movements for social transformation. For me thought and philosophy *begin* there, in the context of the disruption of the status quo, in the desire for radical and transformative practices; and the schemas, ontologies, and frameworks that are produced by philosophy in that context serve as responses to limits in practice or serve to create the space for unheard-of and presently unimagined practices.

Of course, a very generous reading of Derrida might seek to have us understand his work as suggesting much the same point, but it is clear that his political engagements (especially concerning animals) have a very different status in his work than such engagements might have for you and me. After reading through Derrida's remarks on carnophallogocentrism and animals more generally, one can only be left wondering how such concepts and discourse relate to existing practices aimed at transforming our thought and practices concerning animals. If we are seeking to critique ideologies and prejudices while building connections across movements for social transformation, then carnophallogocentrism might be a very helpful concept for such projects—or it might not; we would have to construct and develop the concept and then put it to work within specific contexts. We would have to ask: Does this concept create linkages where none previously existed? Does it allow us to see connections among oppressions that would have gone otherwise overlooked? Does it open up new perspectives and practices? Does it transform and enrich life and thought? Such matters are far more important to me than whether a given concept accurately captures the dominant schema of metaphysics. To that end, I would suggest that the critique of *anthropocentrism* (understood very broadly) might actually do more philosophical, critical, and political work than carnophallogocentrism, as it has the potential (if understood in a very specific and refined manner) to tie together and mutually inform multiple critical analyses, frameworks, and movements for social change. But the main point I would make here is that the creation of concepts has everything to do with creating the space for resistance, transformation, and new ways of living and very little to do with an intra-philosophical, intra-theoretical analysis that hedges every time it is confronted with questions concerning practice and the invention of new forms of life.

Returning to your remarks above, these points about the critical limitations of carnophallogocentrism take on a very direct relevance when they are placed in the context of your comments on the bumper sticker: "Eat Beef. The West Wasn't Won on Salad." You suggest, rightly, that such a statement exemplifies the attitude of carnivorous virility and is simultaneously engaged in "putting down foods associated with women, elevating animal foods, and at its heart, celebrating the genocide of Native Americans." While the concept of carnophallogocentrism would do much to help us make sense of what is going on here, I wonder if it (or any of the other of the myriad discourses on animal ethics) can do full justice to questions concerning, for example, Native Americans and other indigenous peoples and how they figure not just in the constitution of the carnophallogocentric subject but how they figure in and alongside struggles for animal defense. The respective political strategies, epistemologies, and worldviews of the many indigenous struggles for justice and similar movements for animals do not always align, and it seems essential to me for animal theorists and activists to discuss these differences more carefully.

These questions take on a different but still very direct relevance when we think about Derrida's larger strategy in focusing on carnophallogocentrism and anthropocentrism. If these ideologies and practices constitute the dominant schema of metaphysics, then we can be prepared for Derrida to demonstrate how *différance* and other forms of non-presence (ex-propriation and so forth) figure at the very center of the carnophallogocentric subject (and this is what he is doing throughout much of his work on animals). For Derrida, it is in the exploration and thought of how *différance* is at work throughout life that we might begin to challenge the hegemony of presence and its quintessential figure: the carnophallogocentric subject. But the details, implications, and stakes of such a strategy are very rarely discussed by Derrida or his followers. Instead, there is a kind of unspoken faith among Derrideans that there is some direct line between: (1) uncovering and contesting the basic workings of metaphysics, (2) a thought and practice of *différance*, and (3) radical social transformation. To my mind, much more would need to be said here; and I am not all convinced that this kind of approach constitutes a good strategy or a viable ontology.

The Absent Referent II

Now, I would like to turn to some of the more positive and fecund connections between your work and Derrida's work, starting with the issue of the absent referent. Your invention of this concept in the context of issues concerning animals has always struck me as a singularly important achievement. Standard philosophical discussions of animals, especially those deriving from analytic animal ethics, pay lip service (at best) to the ways in which the unique lives of animals and the violence that many animals undergo at our hands is kept from out of sight for certain populations. And yet, there can be little doubt that it is by way of a profound relationship to singular, irreplaceable animals and the violence done to them that many people are moved to transform their lives with regard to animals. Far too little attention is paid to such connections in analytic animal ethics, where a premium is placed on the supposed transformative force of establishing formal, abstract similarities among 'moral patients' and a concern for impersonal justice.

In paying attention to (1) the singularity of the animal beings we encounter and with whom we relate, and (2) to the hidden violence that characterizes the lives of many of those animals, you present an ethical approach that is extraordinarily close in spirit to the logic (if not the letter) of Emmanuel Levinas's ethical thinking and Derrida's appropriation and reworking of Levinas's ethics concerning animals in *The Animal That Therefore I Am*. Levinas argues (contra analytic approaches to ethics) that ethics has its origins in a unsettling, traumatic encounter with the 'face' of the Other ('face' can be understood here, roughly, as any site wherein one encounters the Other's fundamental vulnerability). When I encounter the face of the Other, that Other presents her- or himself outside of the general category of 'others' as a concrete, singular, irreplaceable Other. In such an encounter, I encounter *this* finite Other (as opposed to some abstract 'other' with whom I have formal similarities), and I begin to catch sight of the hidden violence that my egoistic, unthinking existence often entails.

Although Levinas presents his general account of the ethical encounter within the context of inter-human ethics, it has been argued by many of Levinas's best readers that there is no legitimate reason for limiting the 'logic' of this account to human beings alone. One could, in fact, read Derrida's work on animals as a subtle reworking of this Levinasian logic, as an attempt to stretch it and extend it beyond the particular anthro- and androcentric dogmas that plague Levinas's work. If one reads Derrida in this way, then what his work offers us is an attempt to attend to the singularity of animals in their lives and deaths, and joys and sufferings. That all of these things go missing in standard philosophical and metaphysical notions of animals and animality is evidence of the ways in which individual animals are reduced to absent referents in philosophical discourse.

The Iterability of the Carnophallogocentric Subject

Another point of contact between your thought and Derrida's (as well as Judith Butler's) has to do with what you describe as the "reiterative nature of the sexual politics of meat." The carnophallogocentric subject, both in its form as an ideal subject position and as it becomes actualized in individual subjects, is never achieved once and for all. It must be repeatedly enacted, called into being in line with the conceptual-discursive-institutional ideal it invokes (this is what Judith Butler, in her reworking of certain aspects of speech theory, refers to as the performative nature of subjectivity). Thus, as you note, in order to achieve subjective stability it is necessary to find ever new methods and means for reinforcing this identity and shoring it up against that which would unravel it or challenge its dominance.

This notion of the reiterative nature of carnophallogocentric subjectivity has two important implications. On the one hand, it implies that there are the ever new ways of reinforcing carnivorous virility that we need to attend to. On the other hand, the very fact *that* carnophallogocentrism *has to be reiterated* means that it is unstable, structurally open to being challenged and contested, and that it has not fully determined or suffused the various systems of meaning and institutions that constitute individuals as subjects.

Carol:
The need to establish manliness through meat eating has always suggested an instability to masculinity. The difference between 1990 when my book first came out and now is this notion of re-upping or renewing one's "man card." This recent development shows just how unstable masculine identity is perceived to be. (An ad for beer that gives 'man points' for putting together a barbecue but takes away more 'man points' for cooking tofu on it.) Unlike my library card which does not have to be renewed, this 'man card' apparently is constantly being depleted, exhausted, needing re-iteration. So, a strong and very powerful reality—masculine subjectivity—is continually being reinstated through both traditional and new means, while its instability is acknowledged.

Does the space open within its fissures to reconfigure it? In 2013, *Vanity Fair* carried an article ("Steak Shows Its Muscle") celebrating steak as "the butch foodie communion" not just for "flinty-eyed, Armani-suited leaner-than-thou businessmen, but for metrosexuals who wish to beef up their cultural testosterone." A. A. Gill continues: "What does steak say to us and about us? Well, it's manly. If food came with gender appellations, steak would definitely be at the top of the bloke column. Women can eat it, they can appreciate it, but it's like girls chugging pints of beer and then burping. It's a cross-gender impersonation" (2013). The space burps opens; the space closes.

The issue of agency in and against such re-iterative moves also comes up in the way that some vegans who are men appear to accept the givenness of a culture invested in the sexual politics of meat. Trying to show that men are not 'wimps,' and instead can be 'plant-strong,' a few well-known 'manly' vegans coined the term 'hegans.' Rather than pushing into the fissure and suggesting veganism liberates the gender binary, these vegan advocates appear to want veganism to fit into the humanist project, offering the assurance *you are not really changing as radically as you think you are if you become a vegan.* I see it as a conservative response to the threat to the carnophallogocentric subject posed by veganism. To me this seems like they are reassuring what Derrida called the phallogocentricism inherent in Western subjectivity.

It becomes necessary to track the various ways that the subject continues old kinds of exclusions, but has to create new alliances to do this. One of the aspects of these recent meat re-iterations is that they sweat with misogyny. Making connections between my first point ('men need meat') and my second (the function of the absent referent is to hide and promulgate violence), I have started calling some of the representations that have appeared *hate speech* as they celebrate the consuming of the full-bodied female body and position the female body in ways that announce she desires to be consumed. This hate speech normalizes violence.

The Logic of Logocentrism

Matt:
In order to explore further some of the limits characteristic of analytic ethical approaches to animal issues, I thought we might return for a moment to the theme of logocentrism. Above, I glossed Derrida's definition of logocentrism as "denoting the privileges and priorities granted by Western philosophy to the rational, self-aware, self-present, speaking subject." With that gloss in mind, we can note by implication that what gets subordinated by logocentrism are all of those 'things' that fall outside of the *logos*, starting with 'writing' (understood narrowly as written texts and broadly as those things that escape the full control of the sovereign, speaking subject), and extending to all of the other traits (for example, the emotions, passions) and beings (for example, animals, children, nature) that/whom fall short of exemplifying full presence and full *logos*. In his early writings on logocentrism, Derrida sought primarily to make the point that all attempts at achieving logocentric closure and full presence are haunted by *différance*, writing, non-presence, and so on; and in his later writings on animals, he extends this analysis by showing the ways in which logocentrism subtly persists in certain forms of animal rights discourse and politics and produces a contradiction in the reinforcement of the very concept of rights and of the human it seeks to overthrow. I know that you have made related points in your critiques of analytic animal ethics.

Carol:
I like to think that understanding logocentrism helps illuminate why reactions to vegans and vegetarians are often so irate. They have threatened something thought to be *essential* to the subject. It also explains why the defensive

flesh eater's response is to *argue*, to try to defeat through words/arguments the vegan/vegetarian (who is demurring from a culture wordlessly through a dietary change). The non-speaking vegan's dissent must be lifted into the speaking world and there defeated. In *The Sexual Politics of Meat*, I propose "At a dinner where meat is eaten, the vegetarian must lose control of the conversation. The function of the absent referent must be kept absent especially when incarnated on the platter at the table. The flesh and words about it must be kept separate" (2010, 127).

I devote two chapters of *Living Among Meat Eaters* to the problem of talking with meat eaters. I argue that vegans/vegetarians must learn how to stop the conversation, that is, they must refuse to be a speaking subject, refuse to engage at that level. In this, I disagree with the belief among activists (and the almost good-natured assumptions of the average vegan) that we should always answer questions being posed to us. The presumption is that if we are 'the best speaking subject,' that is, if we can proffer forth the best arguments, we will win. I state that conversations "are functioning differently for meat eaters than for vegetarians." And that this logocentric interaction through words, through the speaking self, is "the most stubborn way that meat eaters hold on to their lifestyle" (2009, 91).

In believing they should respond with arguments and explanations of a vegan diet, individual vegans recapitulate the premise and activities of many of the major animal activist organizations. These presumptions include:

1. It is by argument that people change.
2. A debate has a hierarchy, and your goal is to be on the top.
3. If we have the right 'speech' we will prevail. The best arguments will win.
4. The other animals have no voice in human discussions. We must be 'the voice of the animals.'
5. So, the speaking subjects speaking on behalf of the 'voiceless' have to prove themselves to be the best speaking persons or else we have betrayed the non-speaking animals.

Analytic philosophers like Tom Regan, Peter Singer, Steve Best, and Gary Francione accept the logocentric world view, too. They presume the same kind of space, the same kind of subject, the one who has control, the one who is more 'reasonable.' This worldview presumes that change happens this way. As I suggest in "Post Meat Eating," the animal rights movement is a modernist movement in a postmodern time. As you have said, it is all *part of the legalistic and moralist approach to animals*.

Matt:

The issues you raise here concerning the defense of vegetarianism/veganism are important ones to consider; and, like you, I would suggest that standard philosophical arguments have (at best) a derivative and secondary role to play in this area. After speaking with countless meat eaters about vegetarianism/veganism over the past two decades, and after teaching standard philosophical material on vegetarianism/veganism to thousands of students over the past several years, I am more convinced than ever that philosophical arguments nearly always arrive on the scene too late to have the force that most animal ethicists wish them to have. And even when the arguments are considered rationally persuasive by readers, they rarely seem to have the transformative force with non-vegetarians/non-vegans that philosophers claim. I would suggest that for philosophical arguments to carry any persuasive force on these matters there must *already* be in place a certain set of dispositions, relations, and experiences that attune one to animals and their lives. So, even if one wished to retain a space for philosophical argumentation concerning vegetarianism (and I am not entirely opposed to maintaining such a space), it would seem that the space needs to be reinscribed elsewhere than at the foundations of vegetarianism (which is where philosophers would like to place it).

Likewise, when vegetarians/vegans play the role of the rationally persuasive subject in discussions over eating meat, not only does such a gesture place the speaker back into the very logocentric space of mastery that needs to be called into question; it also problematically reinforces the idea that what is at issue here lies in the domain of reason and argumentation (rather than, say, in the domain of what Levinas calls "the face," or emotions, relations, ethical interruptions, and so on). Arguments with meat eaters about vegetarianism/veganism are fairly easy to have, and perhaps even fairly easy to 'win' for the masterful subject, but they rarely bring the discussion into the space where it needs to be in order to get at the heart of the matter—which is, namely, to rethink in a fundamental manner the way one relates and is related to other animals (oneself included).

Another pernicious, but often overlooked, consequence of this subtle reinforcement of logocentrism can be seen in the way that analytic ethics and argumentation map onto larger legal and political strategies for transformation. Given the premium placed on the rational, speaking subject within our logocentric culture, it comes as little surprise that the nonhuman beings animal rightists/welfarists seek to bring into the legal and political sphere most often resemble that same logocentric subject. Animals who can communicate in ways 'we' can understand are more valued than those who cannot; animals who demonstrate 'superior' (which is to say, anthropomorphic) intelligence are considered paradigm examples of animals with moral standing; animals who lack reflexive consciousness, language, familial relations, who are aesthetically disgusting to 'us,' or are culturally unpopular are consistently given less attention in political and legal struggles for animal justice.

In a related vein, and following the same logocentric logic, many legally— and philosophically-inclined animal rightists seek to distance themselves from environmental struggles for justice for nonanimal beings, systems, and regions. It is assumed by nearly every mainstream philosophical and legal theorist for animal rights/welfare that the nonanimal natural world is owed no direct consideration and always and everywhere counts less than humans and animals. And the reasons given for the priority granted to animals are almost always logocentric in nature. Steven Wise's *Drawing the Line* is a prime example of this kind of tendency to exalt logocentric-type animals at the expense of less logocentric-type animals and the rest of the natural world. One of the primary motivations I have for entering standard philosophical and legal debates over animal ethics is to contest these kinds of logocentric consequences; and when I refer to the need for a deconstruction of vegetarianism/veganism[3] and its associated mainstream practitioners, it is precisely these kinds of logocentric limits and blindspots that I believe are in urgent need of deconstruction.

The Power of Phallogocentrism

Carol:
Animal activism not only incorporates the dominant presumptions about the speaking subject, it also operates largely from a *phallo*gocentric position.[4] Both analytic philosophy that argues on behalf of animals and activism prefer the rational, reasonable male speaking voice.

The disowning of the female speaking subject has a long history in the West. But it is one thing to encounter Mrs. Slipslop in Fielding, Mrs. Malaprop in Sheridan, or Tabitha Bramble in Smollett. (Gilbert and Gubar 1979, 30–31). It is another to recognize that animal activism not only privileges the male speaking voice but actively disowns the female speaking subject. I gesture toward the issue of the speaking subject in the second section of *The Sexual Politics of Meat* ("From the Belly of Zeus") which is framed by the story of Zeus's

---

3 "Deconstruction Is not Vegetarianism: Humanism, Subjectivity, and Animal Ethics," *Continental Philosophy Review* 37 (2004): 175–201.
4 See my "Sexual Inequality and the Animal Movement," in Sonbanmastu.

swallowing of Metis, and Zeus's claim that she "gave him counsel from inside his belly" (Adams 2010, 133).

What does animal activism do with women's speaking voice? Their moves are not literally as anthropophagic as Zeus's, but symbolically, they are equally devastating, they announce the animal movement is no longer "just little old ladies in tennis shoes."[5] This comment has been around for decades, but most recently could be found in a profile of Wayne Pacelle head of the Humane Society of the United States, in the *New York Times Magazine*. "'We aren't a bunch of little old ladies in tennis shoes,' Pacelle says, paraphrasing his mentor Cleveland Amory, an animal rights activist. 'We have cleats on'" (Jones 2008). "Don't look at the aging bodies of women activists," they seem to be telling us. This posturing of the animal activist movement tries to fill the cultural space once occupied by the little old ladies (though it still needs them to do the work but hide the fact that they are doing it and that they are 'old' and female).[6]

Several assumptions operate here:

– They assume they are speaking to the dominant subjectivity in the West, the carnophallogocentric subject.
– They assume this subjectivity has trouble/resistance to hearing little old ladies.
– They think they have to accept the limitations in perspective imposed by this subject.
– They have to 'save' animal activism from the threat of empowered little old ladies.

---

5 It might be interesting to think for a moment about why 'little old ladies' have been wearing tennis shoes for so many years. Tennis shoes are certainly better for one's spine than heels; the additional weight that comes with pregnancy results in the widening of the feet; tennis shoes are very comfortable.

6 Animal Studies has been challenged for making this same sort of move resulting in the disappearance of feminist writers who pioneered intersectional theory that included animals and offered early analyses of animal oppression. Susan Fraiman analyzes the disappearance of women in the story of the birth of Animal Studies in "Pussy Panic versus Liking Animals: Tracking Gender in Animal Studies." *Critical Inquiry* 39.1 (Autumn 2012): 89–115. I appreciate her role in asserting the historical and theoretical importance of books like *The Sexual Politics of Meat* and those of my feminist colleagues. The important move is not to accept the either/or assumption presented by some of the gatekeepers of Animals Studies: either English-speaking feminist writers *or* Continental philosophy. This accepts the gatekeepers' formulations by reversal. I believe our conversation shows another way, as does *The Feminist Care Tradition in Animal Ethics*, in which Josephine Donovan and I placed Derrida within the feminist care tradition (2007, 14–15).

- They have to reiterate their rationality over against the stereotype of the emotionally-laden, female-identified body.
- They believe that the *carn* can be excised from the carnophallogocentric subject, plucked out, removed, ruptured, while leaving the phallogocentric subject intact.

Perhaps they believe all this, because this has worked for them.

Certainly the animal activist organizations that display nude and nearly nude photographs of women in their outreach on behalf of animals enact these assumptions. Derrida's concept provides a tool for explaining just why women's naked bodies are so important to some animal campaigns. They think by assuaging phallogocentric subjectivity they can convince him (they clearly are appealing to heterosexual men in much of this) to stop eating meat. They want to remove the *carn* but leave the phallogocentric subject undisturbed. Derrida says in an interview included in *Acts of Literature*, "although phallocentrism and logocentrism are indissociable, the stresses can lie more here or there according to the case; the force and the trajectory of the mediations can be different.... [a] radical dissociation between the two motifs cannot be made in all rigor. Phallogocentrism is one single thing, even if it is an articulated thing which calls for different strategies" (2009, 59–60). Once he appended *carno* to his idea of the subject, did he not also recognize its indissociability from the other parts?

Strategies that assume a culture invested in the carnophallogocentric subject requires the strengthening of the phallogocentric subject as we eliminate the 'carno' aspect (meat eating) like some feminist discourse in Derrida's perspective, "risks reproducing very crudely the very thing which it purports to be criticizing" (Ibid, 60).

Matt:

The quotation from Pacelle is a particularly illustrative example of the standard logic of phallogocentrism that dominates leading forms of animal rights/welfare today. The privileged, 'proper' forms of activism are those carried out by cool, level-headed, rational subjects who believe in the power of arguments and legislation. Women can certainly accede to this privileged space on occasion, but they do so only inasmuch as they renounce all non-logocentric traits, strategies, and considerations. My own experiences with fellow animal activists have suggested to me that precisely the opposite is often the case, that 'little old ladies' are among some of the most important, inventive, and remarkable activists in animal defense circles. And the same holds true for the role of both older and younger women in several kinds of related struggles for social justice, ranging from environmental justice to queer politics to indigenous politics. Women, young and old, employing tactics that fall well outside the logics of logo- and phallogocentrism, have advanced these struggles in unprecedented and vitally important ways.

Returning to the specific context of animal defense politics, I want to underscore that this kind of phallogocentrism also functions to exclude a wide range of additional strategies and activists beyond those just mentioned. We should note, for instance, how a large number of mainstream philosophical and legal animal rightists routinely denigrate direct-action groups like the A.L.F. (and it is significant that younger and older women play a leading role in many of these actions). Such direct-action strategies and tactics that seek to short-circuit the long-term, incremental process of legislating our way to animal rights are often dismissed by mainstream animal rights activists for being not just ineffective (which is a questionable criticism, given the general ineffectiveness of nearly all proposed strategies to date) but also for being driven by many of logocentrism's 'others' (blind emotions, irrational spontaneism, misguided fanaticism, and so on). This kind of dismissal occurs despite the fact that direct-action activists and theorists have developed an extraordinarily insightful series of strategies, practices, analyses, and alternative ways of living in view of animal justice, even as mainstream organizations and theorists leave largely unchallenged the hegemony of phallogocentrism, capitalism, and consumerism in our culture (as your remarks above make clear).

One of the helpful aspects of Derrida's concept of carnophallogocentrism and your notion of the sexual politics of meat are that they help us to attend to these often invisible constraints that guide and limit thought and practice. Did Derrida notice the indissociability of carnivorism with phallogocentrism? Based on his scattered remarks on the issue, one can only conclude in the affirmative. He seems to want to make this series of centrisms not only indissociable but also central to understanding the dominant modes of constituting subjectivity.

This returns us, though, to the question of whether this series (carno-, phallo-, and logo-centrism) is meant to be descriptively exhaustive or only partial, contingent, and strategic. And even if we decide that carnophallogocentrism only functions in the latter sense and does not seek closure over and against other critical analyses of subjectivity, this does not put an end to a whole series of very difficult questions that might arise here. To tease out a bit more one of the issues I raised above: What are we to make of decolonial struggles for social justice that make heavy use of the rhetoric and political strategies of humanism, human rights, and human dignity? How do we link our struggles with theirs when the respective strategies, rhetoric, and histories might conflict? I do not think there are any easy answers to such questions, but I should note here that it is at least clear to me that the resources for working through these matters are *not* to be found in animal defense circles that rely on traditional phallogocentric concepts and practices.

Carol:
Yes, I agree with you completely. It is as though there is a tendency to an anthropomorphic notion of political change.

I like the term *carnophallogocentrism* precisely for what it accomplishes: the linking of carnivorous virility with the speaking subject, and the linking of the Western subject with meat eating.

The carnophallogocentric subject is granted privilege, and this privilege is experienced as pleasure. When this happens the privilege disappears as a social construction and is seen as something private, something personal: "This is your choice, not to eat meat, and my choice is to eat meat." At the minimum, the carnophallogocentric subject is the subject for whom this privilege is working.

So, I find the concept important as it helps to get at the problem of the person who admits to 'carnivorous virility' but who does not want to believe 'he' needs to change. And often the 'virility' part is hidden, it is the naked 'carnivorousness' that is claimed, but it is claimed in an implicitly virile way. The medium becomes a part of the message.

People often respond to *The Sexual Politics of Meat* by suggesting that the phenomenon I am examining is something that is out 'there,' just advertisements, as though they are not implicated in and by it. (*The Sexual Politics of Meat Slide Show* in a sense defeats a part of my purpose because it causes people to think my analysis is about images not attitudes.) Or the response is that *The Sexual Politics of Meat* is critiquing something that has passed (recent advertisements and attitudes notwithstanding). Or that my analysis of images is wrong because how images work has changed. And here is Derrida, in coining the term *carnophallogocentrism*, saying it is about the kind of subject we are, and my point is that this subject is constructed and inflected by a culture heavily committed to the sexual politics of meat.

References

Adams, Carol J. *The Sexual Politics of Meat: A Feminist-Vegetarian Critical Theory.* New York and London: Bloomsbury. 20th anniversary edition, 2010.

Adams, Carol J. *Neither Man nor Beast: Feminism and the Defense of Animals.* New York: Continuum International. Lantern Books Reprint, 2015.

Adams, Carol J. *The Pornography of Meat*. New York: Continuum International. Lantern Books Reprint, 2003/2015.

Adams, Carol J. "Post Meat Eating." In *Animal Encounters*, edited by Tom Tyler and Manuela Rossini. Leiden and Boston: Brill, 2009.

Adams, Carol J. "After MacKinnon: Sexual Inequality in the Animal Movement." In *Animal Liberation and Critical Theory*, edited by John Sanbonmatsu and Renzo Llorente. Rowman and Littlefield, 2011.

Donovan, Josephine and Carol J. Adams. *The Feminist Care Tradition in Animal Ethics: A Reader*. New York and London: Columbia University Press, 2007.

Calarco, Matthew. "Deconstruction Is not Vegetarianism: Humanism, Subjectivity, and Animal Ethics." *Continental Philosophy Review* 37: 175–201, 2004.

Derrida, Jacques. "'Eating Well,' or the Calculation of the Subject: An Interview with Jacques Derrida." In *Who Comes After the Subject?* edited by Eduardo Cadava, Peter Connor, and Jean-Luc Nancy. London: Routledge, 1991.

Derrida, Jacques. "'This Strange Institution Called Literature': An Interview with Jacques Derrida." In *Acts of Literature*, edited by Derek Attridge. London: Routledge, 2002.

Derrida, Jacques. *The Beast & the Sovereign, volume 1*, edited by Michel Lisse, Marie-Louise. 2009.

Fraiman, Susan. 2012. "Pussy Panic versus Liking Animals: Tracking Gender in Animal Studies." *Critical Inquiry* 39.1 (Autumn): 89–115.

Gilbert, Sandra M. and Susan Gubar. 1979. *The Madwoman in the Attic: The Woman Writer and the Nineteenth-Century Literary Imagination*. New Haven and London: Yale University Press.

Hotchner, A. E. 2013. "Steak Shows Its Muscle," *Vanity Fair* (May). Available at http://www.vanityfair.com/culture/2013/05/aa-gill-bull-blood-steak (accessed January 31, 2015).

Jones, Maggie. 2008. "The Barnyard Strategist." *The New York Times Magazine* October 24.

Joy, Melanie. 2010. *Why We Love Dogs, Eat Pigs, and Wear Cows: An Introduction to Carnism*. Massachusetts: Conari Press.

Parry, Jovian. 2010. "Gender and Slaughter in Popular Gastronomy." *Feminism & Psychology* 20.3 (2010): 381–396.

Stevenson, Seth. 2006. "Original SUVs for Hippies? Hummer Courts the Tofu Set." Slate, posted August 14. http://www.slate.com/articles/business/ad_report_card/2006/08/suvs_for_hippies.html Accessed June 13, 2015.

CHAPTER 3

Rotten to the Bone: Discourses of Contamination and Purity in the European Horsemeat Scandal

Nik Taylor and Jordan McKenzie

Introduction

In early 2013 Europe was subject to a moral panic regarding the presence of horse flesh in beef products. Posited as a 'scandal' and often referred to in the media as 'the horsemeat scandal', this topic amassed a large amount of media reports and commentary as well as eliciting police investigations, raids and arrests along with a UK-wide survey of authenticity by the British food safety watchdog, the Food Standards Agency. At the time of writing (and the 'scandal' continues) 'mislabeled' meat products have been found in the Republic of Ireland, the United Kingdom (UK), France, Norway, Austria, Switzerland, Sweden and Germany. Commentators have argued that the 'scandal' demonstrates the complexity of the food industry's supply chains in a globalized world while others have been quick to claim that neo-liberal, free market profiteers are to blame (Hutton 2013).

However, what is strikingly absent in the overwhelming majority of the commentary is any critical discussion of *meat eating* per se. Rather, the opposite is true; by its very absence meat eating is assumed to be normal and is thus further *normalized*. In lieu of any critical engagement with meat eating practices, discourses of contamination, xenophobia and nationhood pervade the coverage of the topic. Analyzing approximately 100 reports carried in the popular media from January to March 2013, this chapter considers the meaning of such a focus and, in particular, why meat eating and animal bodies are the 'absent referent' throughout the discourse. We argue that the media frenzy regarding horse meat is not about health concerns, and is certainly not about animal welfare in general (although we accept that there are speciesist tensions here and that the debate touches upon certain animals' welfare—those we consider worthy, e.g. as 'pets'), but is instead an example of the maintenance of hegemonic species boundaries. In so arguing, we show how binaries between clean and dirty or moral and immoral serve to reinforce the legitimacy of so-called alternatives that are already within the normative standards of a culture. Arguing that the disruption of such normative ideologies and assumptions is how they are made visible, we demonstrate that concern over the supposed pollutant of horse meat in cattle meat is not governed by a fear of disease or contamination per se; rather it is about attempts to reinforce species hierarchies through a normalization of the cultural practice of eating some animals, and not others. We do this by drawing on Mary Douglas's *Purity and Danger* (1966) and through a critical consideration of mainstream anthropocentrist and carnist discourse.

Background to the Horsemeat Scandal

Who would voluntarily eat horse meat? This question conjures up a range of less-than-desirable stereotypes about the 'backwaters' of Europe and the culturally 'backwards' and 'primitive' rural parts of Asia and the Middle East (see Harris 1985 for a historical overview of the consumption of horse meat). Although there are a number of countries where horse meat is consumed much like beef or any other meat, in Australia, northern America and the UK horse meat is commonly found only in pet food and on the occasional risqué fine dining menu. In the European horsemeat contamination scandal in 2013 consumers were outraged not simply about being deceived by food providers, but about the assumed (as far as we know this was never empirically investigated) morally repulsive matter of eating horse. From a sociological point of view there are a number of fascinating phenomena taking place in this hullabaloo, and it is the opportunity to consider some of these that led us to write the current chapter.

Historically, the consumption of meat products in the West was linked to both wealth and prestige, with meat being consumed regularly only by those in the wealthiest classes of society (see, e.g. Cudworth 2011; Fiddes 1991; Vialles 1994). However, as Fiddes reminds us (1991, 166) "to accept unquestioningly that meat is desirable or prestigious, as if these attributes were in some way *inherent* in the substance itself, is to oversimplify the range of ideas that meat supports" [italics in original]. Some of the ideas that meat consumption supports include: that masculinity and virility are dependent upon its ongoing consumption (e.g., Adams 2010); that it is 'naturally healthy' and an essential part of Western economies (e.g., Fitzgerald and Taylor 2014); and, that the meat we consume is a part of a natural cycle wherein animals live among rolling pastures prior to being killed humanely (e.g., Molloy 2011). Scholars have pointed out that behind these 'carnist' ideas (Weitzenfeld and Joy 2014) sits both a profound anthropocentrism and a belief in our ability to control the natural world: "To the many human cultures which have striven to establish their identity apart from and above the rest of nature the consumption of animal meat is an eminently suitable choice to represent power, achievement, prestige, civilization: humanity" (Fiddes 1991, 174). This link between meat eating (emblematic of the control of nature) and civility (see, e.g. Elias 1978, 99; Fitzgerald and Taylor 2014, 167) is still prevalent today but, as fits our globalized world, it plays out at the level of nations and often, as in the current case, is based upon discourses of nationalism.

The examples cited above both create, and rest upon, the assumed normalness of meat consumption. Unless we happen to have a vegan at the dinner table (see Salih 2014), meat eating is considered to be a normal state of affairs and those who avoid it for whatever reason are seen as the oddity. This is especially the case for ethical vegetarians and vegans who are marginalized at best, and harangued and excluded at worse (see Cole and Morgan 2011, for more on this). For the most part this is because "speciesism is woven into our mental, social, and economic machinery, and reproduced through the interaction of these parts—it is a structural aspect of our political-economic order" (Torres 2007, 9). As such the assumed 'naturalness' and 'rightness' of our meat eating practices can be seen as ideological, as "meaning in the service of power" (Thompson 1990, 7). And here, we mean both symbolic power, as in the human control of nature, and material power, as is seen in the fiscal power of the vertically integrated meat processing industry (Cudworth 2011). It is only when such normative ideologies and assumptions are contested that they are acknowledged: rupturing them makes them visible. And this is, we argue, precisely what happened during the horsemeat 'scandal' in Europe in 2013.

There are various tensions evident in the coverage of the 'horsemeat scandal,' and, to a degree they represent broader sensibilities in journalistic coverage of animal issues. For the most part, when animal issues are covered by the media they are the subject of 'soft' news, seen as an antidote to 'hard' news (Molloy 2011, 2). Such coverage rarely, if at all, considers the material reality of animal lives and deaths. As Taylor (2014, 46) makes clear, the media portrays the interests of the corporatized state, and as a result "It is therefore unsurprising that we do not see the misery of the vast majority of animals under human servitude as this would present a challenge to the corporate interests which dominate the Animal Industrial Complex." Almiron and Cole (2015, 1) echo this when they point out that 'The ethics of our treatment of other animals, particularly our treatment of non-'wild' animals, is an almost invisible topic in the media" despite the fact that the media is complicit in manufacturing consent for the ongoing, structurally embedded exploitation and oppression of other animals (it is beyond the scope of the current chapter to fully discuss the role of the media in animal oppression; see Almiron, Cole and Freeman, 2014 for an extended debate). Following this argument, then, it is unsurprising that coverage of the horsemeat scandal, as we shall argue, did little to disrupt normative understandings of human-animal relations and, instead, served to reinforce the notion of human supremacy.

For example, the moralistic rejection of killing horses for meat was driven by two opposing views of the sanctity of the animal. On one hand, horses are constituted as both companions and majestic and sacred animals who deserve a higher status than pigs or cows, and on the other, horse meat is disgusting and primitive—only fit for consumption by the poor, the uncivilized, and our 'pet' dogs who, while they may be loved are also clearly 'less than human.' Meanwhile another unique tension surfaces; why were the people who consider the general consumption of meat to be morally legitimate outraged by the scandal, while those who deliberately avoid meat in their diets were not? In fact, there appeared to be a collective expression of 'who cares?' from the vegan and vegetarian community presumably because they (we) are likely to see the killing of horses for food as *equally unethical* compared to killing other animals. This results in a bizarre state of affairs. It seems that those who defend the position that eating meat is unethical were no more fazed by the scandal regarding the sale and consumption of horse meat than they routinely are by any other 'meat eating' concerns. While those who regularly consume other kinds of meat without guilt or apprehension felt that the practice of consuming horse flesh had to be outlawed. We think this supports our main point. It may seem odd that those who see no, or at least few, problems with eating meat, as evidenced by their regular consumption and defense of it as a practice, were also the ones who most value the black and white distinction between acceptable and unacceptable meat/flesh. However, while accurate, we think this a superficial reading. We believe the consternation felt by 'meat eaters' about the perceived 'contamination' of their meat with horse flesh was a response to their unease at having the normative lines between species contested. On the other hand, (moral) vegans and vegetarians reside in a space where these lines are consistently blurred and thus felt little discomfort (beyond the usual anger and discomfort when faced with casual animal cruelty). This apparent contradiction, and evidence of other tensions in the discourse surrounding the scandal, led us to question what might be going on behind the scenes of the public displays of fear, anger, and outrage at the supposed 'contamination' of 'normal' meat.

One possible explanation is that the emotional attachment that the majority of people feel toward horses is much the same as the moral attachment that vegans and vegetarians feel toward animals more generally. And while this may well be true it is highly unlikely that all of those calling for redress were avid horse admirers, which leads us to conclude that there is something more interesting going on throughout the various discourses surrounding the 'scandal.'

Looking more closely at the media coverage of the topic it starts to become clear that the black and white distinctions drawn between edible and inedible animals served to reinforce the legitimacy of meat consumption rather than challenge it. In itself this is quite a feat and speaks to the power of hegemonic discourse about meat eating: beliefs in the normality of meat eating are so entrenched that even when opportunities arise to question them these threats to 'the norm' are re-framed in such a way to render them toothless. It is with this normalization of meat eating that the current chapter is primarily concerned. Making use of key ideas from Mary Douglas's seminal text *Purity and Danger* (1966) we provide an insight into the moral distinctions of our social norms and show how one seemingly small, and relatively innocuous, debate actually mirrors a much broader social normalization of animal use, meat eating, species difference and human superiority.

Douglas, Purity and Danger

> In chasing dirt, in papering, decorating, tidying, we are not governed by anxiety to escape disease, but are positively re-ordering our environment, making it conform to an idea.
> DOUGLAS 1966, 3

∙ ∙ ∙

> ... an attempt to relate form to function, to make unity of experience.
> DOUGLAS 1966, 3

∙ ∙
∙

Douglas's *Purity and Danger* can be thought of as a key text in the sociology of modernization. Much like Elias's *Civilising Process* (1978) and Durkheim's *Elementary Forms of Religious Life* (1912), Douglas highlights the social drive for rules that implant order on to chaos, conceptions of the civilized on to the uncivilized, and etiquette on to the barbaric. Furthermore, Douglas identifies the significance of distinguishing between the primitive and the sophisticated in the normative process of establishing social standards and ideals. The text is not simply about the distinction between the pure, clean or orderly, and the dangerous, dirty and unpredictable. It is also about the allegedly moral distinction between the characteristics that modernity treasures, and the behavior that it shuns. In this approach, the civilized human is defined in contrast to the primitive other—which could also be read as the distinction between the modern civilization and the natural world. For Douglas, it is through the exaggeration of difference—she uses the example of the gender binary—that "a semblance of order is created" (1966, 5). Consequently, the civilized is defined through the exclusion of the primitive, and the human is defined through the rejection of all that is animalistic.

For Douglas, this is a matter of order and classification, such that anything that is outside or in between classifications constitutes a threat to the moral fiber of modernity. Modern examples of matter that is seen as 'out of place'—of subjects who refuse to conform to a dualized purity—can also be found in the case of bisexual and transgender individuals. Following Douglas's view, these 'unruly' subjects are slighted, oppressed and considered 'out of place' precisely because they disrupt taken for granted categorical distinctions. Those who do not fit neatly and squarely into the binary are seen as problematic and often find themselves labeled, and treated, as deviant. As scholars of deviance have been pointing out for decades (e.g. Becker 1963), this is often a way of demarcating that which we deem important; heterosexuality and the concomitant belief in the normality of two discrete genders in the current example (e.g. Butler 1990).

This point is easily transferred to the human-nonhuman binary. As Steve Baker argues, "the animal is the sign of all that is taken not very seriously in contemporary culture, the sign of that which doesn't really matter" (2001, 174) and in some ways we see that in the media coverage we analyzed in that the animals—the horses themselves—are almost always lost in discussions of contamination, purity, free market trade, consumer confidence in supply chains and so on. But perhaps Baker would have been more accurate if he had argued that animals are a sign of that which does not really matter *unless they transgress sacrosanct boundaries*. At this point they become more visible—at least superficially—although the animals as individual subjects are still likely to be at least partially obscured by generic categorization ('animals', 'horses') and other cultural tropes. Molloy, in her discussion of the seemingly contradictory—sympathetic—tone used by the media to report on animals who escape slaughter (e.g. by leaping 'to freedom' from transport vehicles destined for slaughterhouses) points out that while this kind of coverage *should* rupture the well established binaries between humans and other species that constitute animals as food, it does not. This is because animals are "apolitical and malleable in the sense that they can be made to conform to whatever anthropomorphic devices are used to frame the narrative" (2011, 7). Similarly, while it might seem logical to assume the discourse surrounding the horsemeat 'scandal' should lead to a questioning of our arbitrary demarcation of some species as food and others as companions, it does not. This is because, while the stories are ostensibly about 'consuming horse meat' they actually, through a stress on contamination and purity, reflect rather than challenge dominant values.

Linking Douglas to the Scandal

In many ways Douglas's ideas about the ordering roles of pure binary distinctions pre-figured ecofeminist analyses of the intersections between the oppression of women and of nature which point out that "the maintenance of such dualisms allows for the continued conceptualization of hierarchies in which a theoretically privileged group or way of thinking is superior" (Gruen in Gaard 1993, 80). We heartily agree with ecofeminst arguments that binaries which allow the operation of power in and through them must be dismantled and argue that a starting point is to analyze how such boundaries are constructed and maintained in the first place. We believe an analysis of the various discourses around the horsemeat 'scandal' is an excellent way to achieve this precisely because it is emblematic of the binaristic and hierarchical thinking that underpins the legitimation of meat eating as a cultural practice in general.

In analyzing the various coverage of the 'scandal', it rapidly becomes clear that it is more complex than it first appears and that behind supposed concerns over human health and supply chain contamination lie other deeply held ontological insecurities about the place of humans vis a vis other animals and the control of nature. In this case, the horse meat in question symbolically represents an unclassified and uncertain grey area. One that therefore needs purifying. Read like this, the angst that surrounded the idea of eating horse meat (at this particular cultural nexus) was not one about human health and well-being per se, it was about repairing damage done to our deeply embedded belief system about meat eating. In turn this rests on culturally embedded ideologies regarding what it is to be a civilized human being.

Douglas explains that primitive forms of cleanliness are distinguished from the modern in two key ways: first, that modern cleanliness is a matter of culture rather than religious tradition or superstition, and second, that in the present day we have scientific knowledge of bacteria that is a form of 'dirt' that is invisible to the naked eye (1966, 44). Although both points are interconnected, it is the first that is of particular significance (since horse meat is no more or less dangerous or unhealthy than other kinds of meat). Yet there are religious origins to the cultural norms surrounding edible and inedible meats. The remnants of biblical distinctions (such as the sacred and profane in Durkheim) are never very far from Douglas's evaluation of modernization, yet she appears to identify a trend that is of far greater concern for the present day. The distinctions between sacred and profane exist as though they are handed down from a higher source (even when Durkheim exposes their social origins). The status of this external source functions as a justification for the often cruel and irrational treatment of the other. Yet for Douglas, the distinction between purity and danger is purely a characteristic of society. It is self-regulation applied on a mass scale—much like the self-regulation of meat consumption.

For Douglas, when confronted with an 'anomaly' (such as horse meat) we can either accept it by finding a way to incorporate it into a pre-existing model (or create a new model), or we can reject it as filth; as a threat to order. Here we can think of the Australian restaurateurs and butchers who have been suggesting that we accept horse meat as part of a new culinary genre (*Herald Sun* 06 Feb 13; *Sydney Morning Herald* 9 July 2010). An example of the positive integration of an anomaly can be found in the highly processed and unnatural meat products of the modern era. Why is hotdog meat considered to be more normatively pure and clean than an avid barbecue enthusiast slaughtering his or her own meat? Considering the concerns of bacteria and the consequent use of chemicals in the factory production of processed meat products, there is a great contradiction in the classification of slaughtering by hand as 'dirty' and processed meats as 'pure'. Our point is not that we should start to kill our own meat, but that the classification of purity and danger is not dependent upon fact. Rather it is a cultural stipulation based upon the norms of the time.

Methods

Throughout the remainder of the chapter we analyze selected coverage of the horsemeat 'scandal' as seen in the national press in the UK. We do this following Molloy's (2011, 7) arguments that "Although popular media texts rarely present a direct challenge to the social order, they are nonetheless a crucial aspect of the cultural processes which set discursive limits on topics … [that] … media discourses can shape public understandings of animals in ways that appear to be natural and normal". First, a quick note on the data. This book chapter comes out of a larger project which analyzes the discourse surrounding the coverage of the horsemeat scandal. For this chapter we made use of data collected from British newspapers. Using ProQuest we searched for "horsemeat" and "horsemeat scandal" in all English speaking newspapers between 1st

January 2013 and 31st March 2013. We then filtered the data collected to include only the national British daily newspapers (*The Daily Mail, The Sun, The Times, The Guardian, The Daily Mirror, The Daily Telegraph, The Independent*) and national Sunday newspapers (*Sunday Times, Mail on Sunday, The Independent on Sunday, Sunday Mail, The Observer, The Sunday Telegraph*). The remaining stories were entered into QSR NVivo 10 which we used to thematically code the data.

The themes apparent in the data are multiple and often intersecting. Thus, while we see evidence of meat eating being linked to ideas of control through its link to human superiority, there is also evidence that the discourse reflects a desire to re-establish control over markets, following the global financial crisis (GFC); control over ecosystems and nature, given climate change concerns; and control over national sovereignty given the uneasy political linking of the UK and the EU. While we may touch upon these in the analysis below they are not our focus here. Our focus, for this chapter, remains on what the discourse can tell us about meat eating cultures.

Purity, Taint, and Contamination

In an analysis of the media coverage of the horsemeat scandal, we have identified three central themes which overlap substantially; and can be organized under the headings; purity, taint and contamination. Each of these three dominant themes relate to aspects of *Purity and Danger*, and each contain multiple points of interest. Due to their intersecting nature we discuss them together below.

The media coverage we looked at is replete with the terms 'tainted', 'contamination' and 'contaminated'. The terms occur 182 times in the data set (tainted = 11; contaminated = 81; contamination = 90); by point of comparison the highest frequency was 'horse' with 623 occurrences, the second highest was 'horsemeat' which appeared 312 times. On all occasions the idea of contamination was assumed rather than argued or proven, as with *The Daily Mail* (no author, Feb 25, 2013), whose story titled "The Buck Must Stop with the Food Standards Boss" opened with the words "The contaminated meat scandal...". Occasionally the idea that horse meat is intrinsically a form of contamination is distinctly paired with other alleged forms of taint as in the story from *The Daily Mail* (Lynott, 8 Feb 2013) which asserted that "lasagne is tested for traces of horse drug", or the article in *The Daily Mirror* (Beattie, 28 Jan 2013) which claimed "Revealed: How deadly horse toxins could end up in our food". Similarly *The Guardian* (Lawrence, 25 Jan 2013), with slightly less sensationalism, considers "Horse Meat Carcinogen Found In Food Chain: Drug Came

From Animals Slaughtered In Britain: Contaminated Meat Was Consumed In France." Tainted appears much less frequently than the idea of contamination and is usually restricted to articles which discuss the need, or the provision, of scientific tests to determine meat purity, as in *The Daily Mail* (no author, 1 March 2013) article which claimed "Poland's General Veterinary Inspectorate has admitted it found three tainted samples from 121 tested, with 80 more to be examined." The idea of contamination and taint was often paired with both consumer confidence and scientific testing which was posited as a way to return consumer confidence; and time and again the articles tell us that there is no threat to human health even if the meat is "tainted".[1] An example of this is the statement in *The Daily Mail* article (Barnes, 23 Feb 2013) that quotes a "North Lanarkshire spokesman" as saying "We will continue to carry out additional testing in the coming days. We cannot confirm that these products have not been consumed. However, the consumption of horse meat is not considered harmful to health." The exceptions to this are three articles which discuss the idea that the meat may be tainted from drugs given to elderly, sick, or ex-racing horses.

The question of whether the 'contaminated' meat was harmful to human health overlapped with another clear concern in the coverage: trust. This included both consumer trust in the supermarkets and in the food industry more broadly. Many articles cite consumers who do not wholeheartedly object to consuming horse meat, rather they object to being lied to on the packaging of the meat they purchased presumably thinking it was cow-meat. This theme overlaps with the question of health concerns, as many consumers did not feel that they were put at risk when being fed horse, it was the dishonesty that upset them. However as the story unfolded, journalists found evidence that the formal record keeping procedures required for cow-meat were not being adhered to with horse. Consequently cases eventually surfaced where horse meat contained traces of drugs and medication that could be harmful to consumers. To our knowledge, there were no reported cases of consumers falling ill as a result of consuming horse, and so the industry managed to avoid comparisons to the bovine spongiform encephalopathy (BSE) scandal decades earlier.

Whilst health risks were a possible concern, they were severely outweighed by the matter of trust that consumers lost in supermarket food products. This was often tackled in the articles by promises from various officials regarding

---

1   It is beyond the remit of the current chapter to consider the entangled discourses of scientific knowledge, technocentric ideas about 'fixing' the horsemeat 'scandal' and normative understandings of meat production and consumption. We read this, however, as part of the links between the animal- and pharmaceutical- industrial-complexes (See Twine 2012).

the oversight of supply chain integrity and promising greater transparency in the future. *The Daily Mail* (Murphy, 23 Feb 2013) tells its readers that one of the companies allegedly involved in the scandal states on its website "we work closely with our producers to ensure the quality and integrity of our supply chain." And *The Sun* (Phillips, 7 Feb 2013) quotes a 'spokesman' for Aldi (a supermarket chain) as saying "investigations are continuing. We will continue to maintain active scrutiny across our supply lines and will always put the quality of our products and safety of our customers first".

A different way of addressing public concern and mistrust in the industry is seen in many of the articles which invoke a sense of nationalism and a xenophobic fear of 'foreign' criminals, often at the same time. *The Daily Telegraph* (Blackden, 28 Feb 2013) states "In an effort to win over consumers, Tesco said it would source more of the meat it sells from Britain" thus implying that British sourced meat is safer. Many of the stories point to the 'foreign' ownership of the slaughterhouses and processing plants that were allegedly to blame for the 'contaminated' meat making its way to Britain. Some of this is done covertly by referring continually to suppliers "in Poland" or "in the Czech Republic". More often, however, it is done overtly as with the headline from *The Daily Mail* (Murphy, 23 Feb 2013), "Firm Labelled Horsemeat as Beef—in Czech" even though this is a story about an Irish firm allegedly supplying 'contaminated' meat overseas. Similarly *The Guardian* (Lawrence, 8 Feb 2013) explains "it was 'highly likely' that criminal activity was to blame for the contamination, the Food Standards Agency said". Meanwhile, *The Independent* reported (Hickman & Masters, 8 Feb 2013) the Environment Secretary Owen Paterson as stating "consumers can be confident that we will take whatever action we consider necessary if we discover evidence of criminality or negligence". And relying on both criminality and xenophobia, *The Guardian* (Harding & Traynor, 14 Feb 2013) explains that "The scandal has focused attention on the murky pan-European supply chain for meat products, which stretches from abattoirs to supermarkets via mysterious offshore companies." *The Financial Times* is even more bold in its headline assertion (Carnegy et al., 9 Feb 2013) that "Horsemeat Scandal Linked to Criminals".

These kind of claims are essentially linked to status which invokes simple comparisons (criminal versus law abiding) that mirror the ideas of purity and contamination. Criminals are linked to, and ultimately responsible for, contamination and law abiding consumers are the ones who pay the price by consuming tainted meat. However, invoking status was not limited solely to the idea of criminality. Status included issues relating to nationalism, class, culture and, at times, racism. The media coverage distinguishes adventurous 'foodies' who consume horse as a part of a sophisticated palate, from lower socioeconomic parts of Europe where consuming horse is more of a necessity. Douglas describes this as 'ritual pollution' which arises as an embarrassing failure to meet common standards of "aesthetics, hygiene or etiquette" (2002 [1966], 92) and results in social sanctions such as "contempt, ostracism, gossip and perhaps even police actions" (92).

The accidental consumer of horse meat risks being mistaken for 'primitive' in Douglas's terminology, a status that is applied only to those who are below the status of the term's user. In addition to the status of the consumer, the status of the animal is also called into question. One article suggests other animals are even more controversial than horse, "Concern across Europe grew last night amid claims that unfit minced horse, pony and even DONKEY could have found its way to British shop shelves in a criminal food chain" (Parker, *The Sun* 12 Feb 2013). Although in most circumstances horses are given a higher status than cows, sheep and chickens (as companions, as athletes, as soldiers etc), the status of horse on the plate seems to be a central concern in the media coverage. For Douglas, the most worrying threats to the dichotomies of modernity are the objects that fail to fit within one category or the other. In the present case, this could simply be understood as horses fitting in to the 'animals that are not for eating' category alongside dogs, cats and (in Western cultures) whales. The dramatic response to the discovery of horse meat in products that were supposed to only contain 'animals that are for eating' was due to the challenge to this dichotomy. If the response to consuming horse meat is anything less that dramatic, then which dimension of the dichotomies that hold normative social values will be the next to fall?

That horses are considered to be of a low status, certainly a lower status than humans, and thus could be seen as a potential 'contaminant' is seen in some of the headlines that are presumably supposed to be humorous. *The Sun* (Clarke, 28 Feb 2013) ran the headline "Balls busted . . . now wieners pulled" and again, *The Sun* (no author, 21 Feb 2013) used the headline "It's bully for beef". On 17 Feb 2013 *The Sun* used the idea of "nag-gate drug fear" (Woodhouse) and on 12 Feb 2013 they stated in another headline that "It's a Load of Old Pony" (Parker). *The Daily Mail* (Sears & Steiner, 09 Feb 2013) ran a headline which tells us of a "Tycoon who loves ponies (but not on a plate)", while *The Sun* (Phillips, 7 Feb 2013) tells us "Now it's Nag Bol".

Taken together, these themes create an impression of a 'scandal' that is at one and the same time potentially serious but also a bit of a joke. The seriousness lies in early concerns that horse meat could be harmful to human health as well as concerns about the integrity of the food supply chain across Europe. The humor comes after it has been established that there is little harm to humans expected and the idea that it is 'only horses' and probably 'nags' (usually tied to the idea of the slaughtered horses being ex racing industry). Overall this speaks to a normalization of meat eating (as soon as we allay concerns that it may damage human health, the issue becomes one of being lied to as consumers) that stems from a callous lack of concern for other animals and their welfare. Few of the articles mentioned animal welfare at all, and when they did it was usually linked to ideas of scientific testing of animal meat to ensure its suitability for human consumption. The attempts at humor (presumably) that we outline above demonstrate the apotheosis of this attitude which pervades the articles whether from tabloid or broadsheet newspapers.

Given this lack of focus on meat eating per se, and the—we would argue—deliberate framing of this issue as one distinctly *not* about animal flesh, this lack of attention to animal welfare is unsurprising. Similarly, the fact that discussions about alternatives to animal flesh consumption are largely missing from the coverage is hardly surprising given the can of worms that could potentially open. We did find some evidence of themes regarding the boosting of interest in vegetarian/meat substitute products and waste but to a much lesser degree than others. While one article cited that meat substitute products made by Quorn had increased by 15% it focused on how sales in fish had increased by 20%. This is in keeping with our argument that coverage actually underlines the assumed normality of flesh consumption, here by suggesting that the concerns over 'beef' products were more likely to lead consumers to other meats before changing the meat eating habits entirely. Further, several articles also cited that consumers were increasingly interested in buying local/ British beef, as it was deemed to be more trustworthy (which ties has to the theme of nationalism, status and trust). From this it seems reasonable to conclude that concerns over animal welfare played little to no role in the decisions of consumers.

Finally, worthy of note here is that the theme of food wastage is repeatedly mentioned from an entirely uncritical perspective. In an attempt to recover from the damage to their reputations, supermarkets seemed to boast about the disposal of products that could possibly contain horse. Yet there was no mention of the deplorable waste of food that was not deemed to be unsafe, food that many would surely be happy to take for free. Instead, the disposal of vast quantities of otherwise safe food was used as a technique that distanced supermarkets from the controversy. After disposing of an entire range of products from a particular supplier, one supermarket spokesperson stated "we have no positive test results for horse DNA ... but feel that it is the right thing to do" (Poulter, *The Daily Mail* 15 Feb 2013). Given links between climate change and meat production this seems a particularly irresponsible response (see, e.g. Taylor 2012). It is, however, entirely in keeping with a response designed to assuage fears in consumers and repair any damage done to the belief in the necessity of animal flesh in the modern Western diet.

In short, then, whether it be discussions regarding consumer trust and health, brief considerations of alternatives to 'beef' or assessments of the contaminative potential of horse flesh, pervading all of the coverage we analyzed is a pernicious anthropocentrism, itself linked to, and justified by, carnist discourse (Joy 2001; 2010).

Anthropocentrism and Carnism

Anthropocentrism is a belief system which posits humans as the centre of the universe and then uses this schema to measure everything else against them. For nonhuman animals this means they always, and irrevocably, are found wanting compared to their human 'superiors'. Whether it is because they lack the ability to talk, project into the future, use tools, display complex emotions like empathy or depression, they are held against a human benchmark and found wanting at every turn. Of course, the benchmarks change over time, usually after investigations have 'proven' that nonhuman animals can, indeed, share a particular trait. For example, following work in the 1970s demonstrating that great apes could not only use tools but could master human forms of communication like sign language (e.g. Gardner and Gardner 1969), the goal posts were moved and instead of language being the reason other animals were thought not to deserve equal moral worth, we began to look at morality as the cornerstone of humanity. In turn this has been challenged recently (Bekoff and Pierce 2009) and so we see the benchmark now being empathy or community, or some other form of human construction that animals cannot (at the current time, within current paradigms and using contemporary investigation techniques) hope to meet.

Anthropocentrism therefore goes a long way toward justifying our treatment of nonhuman animals as it justifies their exclusion from our moral circles (at least, this is the story we tell ourselves). Scholars are keen to point out, though, that anthropocentrism is a social construction; as Weitzenfeld and Joy (2014, 3) argue, "anthropocentrism, which has narcissistically privileged humans as the center of all significance, is not an innate disposition but a historical outcome of a distorted humanism in which human freedom is founded upon the unfreedom of human and animal others." As they go on to explain, this is a deeply embedded, structural phenomena, "Anthropocentrism is not the effect of inescapable, ahistorical constraints of human sensibilities, but rather it is a historical development born from specific institutional and philosophical traditions" (2014, 5).

This is important because it means that, as a social practice and the outcome of historical and socio-political contingencies, the operation of anthropocentrism can be identified and deconstructed. In other words, anthropocentrism as a process can be analyzed. One effective way to do this is to consider the carnist discourse prevalent in most [writings] about meat and meat consumption. Weitzenfeld and Joy (2014), drawing on Joy's earlier work (2001; 2010), explain that carnism is a "sub-ideology of speciesism that dichotomizes non-human animals into "edible" and "inedible" categorizations and legitimates the exploitation and consumption of animal others" (21). Much of this is achieved at the point where animals are killed, by making slaughterhouses, slaughterhouse work and workers, invisible and marginalized (see Vialles 1994) but it also occurs in narratives of consumption where "carnistic defenses enable gross cognitive and affective distortions in order for human consumers to support the system" (Weitzenfeld and Joy 2014, 21).

As can be seen in the above examples of coverage of the 'horsemeat' scandal, carnistic and anthropocentric discourses abound. The central concerns are human health and other socio-political issues that affect humanity, such as criminal activity or trust in food supply chains. When animal welfare is mentioned it is in a firmly welfarist discourse which assumes the normalcy of meat consumption. For example, questions are asked about humane killing instead of about the right to kill others for our food in the first place. The anthropocentrism and allied carnism that pervade the coverage are built upon, and reinforce, binary constructions of human and animal which themselves rest upon constructions of contamination and purity.

Concluding Remarks

Throughout this chapter we have demonstrated that the concerns surrounding the 'horsemeat' scandal are more concerns about human-animal binaries than they are about purported issues of safety in consuming horse meat. By linking this to a discussion of Douglas's work on the utility of social constructions of purity and contamination we have argued that the scandal coverage feeds neatly into constructions of anthropocentrist and carnist discourses. In other words, "By naming the belief system which underlies the acts of meat production and consumption we are better able to acknowledge that slaughtering nonhuman animals for human consumption is not a given but a choice; a choice that is based upon an ideology in which the domination and exploitation of other animals is considered a natural human privilege" (Joy 2001).

The matter of the public's use of ethical and moral claims sits in a rather unique position in relation to Douglas's work. For Douglas, the distinction between pure and dangerous, or clean and dirty, is principally a matter of ethical or unethical. To project the label of 'dirty' is not simply a claim regarding cleanliness, it is a moral judgment. This is made most clear in Douglas's analysis of incest as a deeply immoral act that is deemed repulsive predominantly though the act's label as dirty. Therefore, to be on the wrong side of a normative judgment, is not simply a matter of not being normal, it is primarily a matter of morality. The distinction between incest and socially acceptable forms of intimate relationships does not challenge the practice of intimate relationships in general, it tends to further legitimize social norms. And this distinction functions in a similar way in the example of the horsemeat 'scandal'. The moral distinction between eating horse and eating other animals does not pose as a critique of eating meat in general. It fails to raise questions of whether the entire industry is immoral. And in doing so, it further normalizes the practice of using animals for food. Douglas is clear that these normative distinctions generally exist independent of scientific knowledge, and that in many cases the lines drawn between acceptable and unacceptable are entirely arbitrary in a practical sense.

In summation, we have found that the European Horsemeat Scandal of 2013 may have done more to normalize meat consumption than highlight its risks and ethical contradictions. By containing the public's rejection of unknown and unlabeled meat products within the example of horse meat, greater interrogations into matters of whether cows or pigs deserve the same status as horses, were avoided.

References

Adams, Carol J. *The Sexual Politics of Meat: A Feminist-Vegetarian Critical Theory*. New York: Continuum, 2010.

Almiron, Nuria, Cole, Matthew and Freeman, Carrie Packwood. *Critical Media and Animal Studies: Communication for Nonhuman Animal Advocacy*. London: Routledge, 2015.

Baker, Steve. *Picturing the Beast: Animals, Identity, and Representation*. Champaign, Illinois: UP, 2001.

Becker, Howard S. *Outsiders: Studies in the Sociology of Deviance*. New York: The Free Press, 1963.

Bekoff, Marc, and Pierce, Jessica. *Wild Justice: The Moral Lives of Animals*. Chicago: University of Chicago Press, 2009.

Butler, Judith. *Gender Trouble: Feminism and the Subversion of Identity*. London: Routledge, 1990.

Cole, Matthew, and Morgan, Karen. "Vegaphobia: Derogatory Discourses of Veganism and the Reproduction of Speciesism in UK National Newspaper." *British Journal of Sociology* 62.1 (2011): 134–153.

Cudworth, Erika. *Social Lives with Other Animals: Tales of Sex, Death and Love*. London: Palgrave 2011.

Douglas, Mary. *Purity and Danger: An Analysis of Concept of Pollution and Taboo*. London and New York: Routledge, (1966) [2002].

Dunn, Mark. "Melbourne diners chomping at the bit for equine menus." *Herald Sun*, Feb 6, 2013 http://www.heraldsun.com.au/news/victoria/melbourne-diners-chomping-at-the-bit-for-equine-menus/story-e6frf7kx-1226571121976. Accessed 27/02/2014.

Durkheim, Émile. *Elementary Forms of Religious Life*. Translated by Joseph Ward Swain. New York: Free Press, 1912.

Elias, Norbert. *The Civilising Process*. Translated by Edmund Jephcott, Oxford: Blackwell, 1978.

Fiddes, Nick. *Meat: A Natural Symbol*. London: Routledge, 1991.

Fitzgerald, Amy J., and Taylor, Nik. "The Cultural Hegemony of Meat and the Animal Industrial Complex." In *The Rise of Critical Animal Studies: From the Margins to the Centre*, edited by Nik Taylor and Richard Twine, 165–182. London: Routledge, 2014.

Gardner, Allen, and Gardner, Beatrice. "Teaching Sign Language to a Chimpanzee." *Science, New Series* 165(3894), Aug 15 (1969): 664–672.

Gruen, Lori. "Dismantling Oppression: An Analysis of the Connection between Women and Animals." In *Ecofeminism: Women, Animals Nature*, edited by Greta Gaard, 60–90. Philadelphia: Temple University Press, 1993.

Harris, Marvin. *Good to Eat: Riddles of Food and Culture*. New York: Simon and Schuster 1985.

Hutton, Will. "The Meat Scandal Shows All That Is Rotten about Our Free Marketers." *The Guardian*, Feb 13, 2013. http://www.theguardian.com/commentisfree/2013/feb/17/horsemeat-scandal-is-tory-party-crisis. Accessed 10/6/15.

Joy, Melanie. "From Carnivore to Carnist: Liberating the Language of Meat." *Satya*, September, 2001. http://www.satyamag.com/sept01/joy.html.

Joy, Melanie. *Why we Love Dogs, Eat Pigs and Wear Cows: An Introduction to Carnism*. San Francisco: Conari, 2010.

Molloy, Claire. *Popular Media and Animals*. London: Palgrave, 2011.

Pepper, Daile. "Horses for courses as diners eye off equine entrees." July 9, 2010. http://www.smh.com.au/action/printArticle?id=1682092. Accessed 27/02/2014.

Salih, Sarah. "Vegans on the Verge of a Nervous Breakdown." In *The Rise of Critical Animal Studies: From the Margins to the Centre*, edited by Nik Taylor and Richard Twine, 52–68. London. Routledge, 2014.

Taylor, Nik. "Reversing Meat-Eating Culture to Combat Climate Change." Report for *The World Preservation Foundation*, 2012. http://www.worldpreservationfoundation.org/Downloads/ReversingMeatEatingCultureCC_NikTaylor_140612.pdf Accessed 10/6/15.

Taylor, Nik. "Suffering is not Enough: Media depictions of Violence to Other Animals and Social Change." In *Critical Media and Animal Studies: Communication for Nonhuman Animal Advocacy*, edited by Nuria Almiron, Matthew Cole and Carrie Packwood Freeman, 42–55. London: Routledge, 2015.

Thompson, John B. *Ideology and Modern Culture: Critical Social Theory in the Era of Mass Communication*. Cambridge: Cambridge University Press, 1990.

Torres, Bob. *Making a Killing: The Political Economy of Animal Rights*. West Virginia: AK Press, 2007.

Twine, Richard. "Revealing the 'Animal—Industrial Complex'—A Concept & Method for Critical Animal Studies?" *Journal for Critical Animal Studies* 10.1 (2012).

Vialles, Noilie. *Animal to Edible*. Cambridge: Cambridge University Press, 1994.

Weitzenfeld, Adam, and Joy, Melanie. "An Overview of Anthropocentrism, Humanism, and Speciesism in Critical Animal Theory" In *Defining Critical Animal Studies: An Intersectional Social Justice Approach for Liberation*, edited by Anthony Nocella, John Sorenson, Kim Socha and Atsuko Matsuoka, 3–27. New York: Peter Lang Publishing, 2014.

Media Reports

No Author. The buck must stop with food standards boss [Scot Region], *Daily Mail* [London (UK)] 25 Feb 2013: 54.

No Author. Horse DNA is found in beef at three Polish suppliers [Eire Region], *Daily Mail* [London (UK)] 01 Mar 2013: 10.

No Author. It's bully for beef, *The Sun* [London (UK)] 21 Feb 2013: 46.

Barnes, J. School beef gets chop: Councils told to bin supplier's products as horse DNA is discovered in burger [Scot Region], *Daily Mail* [London (UK)] 23 Feb 2013: 6.

Beattie, J. Revealed: How Deadly Horse Toxins Could End Up In Our Food: Ireland Beef Scandal carcasses Had Cancer-Causing Drugs criminal Gangs Control Illegal Trade [Ulster Region], *The Daily Mirror* [London (UK)] 28 Jan 2013: 10.

Blackden, R. Tesco vows to test all meat, but warns move may push up prices. *The Daily Telegraph* [London (UK)] 28 Feb 2013: 5.

Carnegy, H., Lucas, L., Rigby, E., Smyth, J. and Stacey, K. Horsemeat scandal linked to criminals. *Financial Times* [London (UK)] 09 Feb 2013: 3.

Clarke, J. Balls busted. Now wieners pulled: Sausages withdrawn after horsemeat scare [Eire Region], *The Sun* [London (UK)] 28 Feb 2013: 6.

Harding, L., and Traynor, I. Dutch trader could be central player in horsemeat scandal: Middleman 'convicted last year': Products withdrawn across Europe. *The Guardian* [London (UK)] 14 Feb 2013: 1.

Hickman, M., and Masters, S. Beef scandal will lead to new food testing regime. *The Independent* [London (UK)] 08 Feb 2013: 2.

Lawrence, F. Horse meat carcinogen found in food chain: Drug came from animals slaughtered in Britain: Contaminated meat 'was consumed in France', *The Guardian* [London (UK)] 25 Jan 2013: 5.

Lawrence, F. Findus clears shelves of beef lasagne in horsemeat scare: Product contains between 60–100% horsemeat: UK ministers accused of 'being asleep on the job'. *The Guardian* [London (UK)] 08 Feb 2013: 8.

Lynott, L., Lasagne is tested for traces of horse drug: Fears prompt two chains to withdraw products [Eire Region], *Daily Mail* [London (UK)] 08 Feb 2013: 15.

Murphy, S. Firm labelled horsemeat as beef—in Czech: Tipperary abattoir closed down as industry plunged into new crisisA sorry affair that has blighted our farm trade [Eire Region], *Daily Mail* [London (UK)] 23 Feb 2013: 1.

Parker, N. A load of old pony: exclusive: inside horse meat slaughterhouse [Scot Region], *The Sun* [London (UK)] 12 Feb 2013: 16.

Phillips, R. Now it's nag bol: Aldi pulls spaghetti ready-meals over horse meat fear [Eire Region], *The Sun* [London (UK)] 07 Feb 2013: 6.

Poulter, S. Asda withdraws soup over horsemeat fears [Scot Region], *Daily Mail* [London (UK)] 15 Feb 2013: 24.

Sears, N., and Stenier, R. Tycoon who loves ponies (but not on a plate), *Daily Mail* [London (UK)] 09 Feb 2013: 9.

Woodhouse, C. Nag-gate drug fear: exclusive 2 [Edition 2], *The Sun* [London (UK)] 17 Feb 2013: 2.

CHAPTER 4

Live Exports, Animal Advocacy, Race and 'Animal Nationalism'

Jacqueline Dalziell and Dinesh Joseph Wadiwel

In March 2011, an animal advocacy non-government organization, *Animals Australia*, investigated Indonesian abattoirs receiving live exported animals from Australia. The results of the investigation and subsequent footage were released to the general public, depicting graphic acts which were termed 'torture,' such as cows being subjected to "abuse through eye gouging, kicking, tail twisting and tail breaking" prior to slaughter (*Animals Australia* and RSPCA 2011). The exposé led to what has been one of the most successful animal advocacy campaigns in recent Australian history. The allegations of mistreatment in Indonesia, and in other live export destinations, intensely occupied the national press for months (and continue to do so to this day). Public concern spilled over into public protest, when on April 14 2011 mass demonstrations across Australia outside government buildings demanded the Prime Minister call for a conscience vote to ban the trade (SMH 2011). The scale of protests was unprecedented for an animal advocacy campaign in Australia (O'Sullivan 2011). The campaign culminated in increased public scrutiny and government response; the ensuing momentum led to a temporary ban on live exports to Indonesia, Australian government reviews, and changes in practice and regulation.

This chapter poses the question: Why has the campaign been so successful? We might speculate that a growing awareness and sympathy for animal welfare and/or rights arguments has generated the public's current discomfort. However, if growing national concern over animal welfare is premised upon an increased moral recognition of animals, there are some peculiar inconsistencies. An animal advocacy argument for the banning of the live exports trade might be based on at least two factors: that the export journey abroad represents an additional cruelty animals would not have met if slaughtered in Australia; and that the animal welfare standards of destination countries are not as rigorous as Australian standards, thus animals invariably undergo cruelty they otherwise would not. Yet both arguments rely on assumptions that are contestable. Firstly, live export animals are not the only animals that are transported long distances, given animals are regularly transported immense distances *within* Australia as part of the routine domestic agricultural industry (WSPA 2008, 29). Why has public attention been focused on live exports by sea, as opposed to the welfare issues of animal transport in general, particularly across the vast internal distribution networks in Australia? The second point upon which the debate pivots is the unquestioned belief that welfare standards abroad are worse than those within Australia. The public response was aroused precisely by instances of cruelty beyond Australia; cruelty that has been assumed is representative of slaughter practices in export destinations. Although widespread cruelty has also been exposed, as we shall discuss in more detail below, in a range of Australian abattoirs. There has been concentrated public concern in relation to the use of 'traditional' forms of killing at export destinations, where animals are not stunned prior to being slaughtered. However, abattoirs within Australia also practice forms of Halal or Kosher killing, many of which utilize stunning prior to slaughter, but in some cases approval is provided for no prior stunning (Puddy 2011; RSPCA 2015). If the public concern has been in part due to the use of 'religious' killing practices within abattoirs in export destinations, why is it that identical practices in Australia have not aroused the same concern? We suggest that these contradictory threads within the live exports campaign warrant closer attention.

In this chapter we focus on the way in which the current discussion of live exports of animals and the focus on the welfare of slaughter practices in export destinations might occur within the context of a racialized geopolitics. We acknowledge upfront the difficulties associated with raising these questions. We do not support continued, systemic violence against animals—including animal containment, transport, breeding, or slaughter for human consumption—wherever this occurs. Crucially, we do not aim to condense the importance, nor urgency, of the valuable advocacy done by activists and organizations working to eliminate this violence. However, within the context of a broader consideration of strategy in confronting systems of violence against animals, we seek in this chapter to map the racialized terrain within which animal advocacy inevitably occurs. Specifically, we seek to understand the potential for campaigning for animal welfare and rights to in turn simultaneously enact forms of violence and marginalization, particularly against racialized humans. We pose these questions amidst work by other scholars which have probed in similar directions, including Amy Breeze Harper (2010), Claire Jean Kim (2007, 2010, 2015), Maneesha Deckha (2012), and Will Kymlicka and Sue Donaldson (2014).[1]

---

[1] This chapter also builds on a shorter discussion of live exports in Dinesh Joseph Wadiwel, *The War against Animals*.

Visibility and Animal Abuse

A key feature of the live exports campaign has been the use of video footage to broadcast acts of cruelty. Visibility is paramount here: the increased availability of video footage—and the decisive role of emerging information technologies in enabling the capture of acts of cruelty via small devices such as mobile phones—facilitates mass and potentially 'viral' access to the general public. Given the wider availability of footage depicting cruelty towards animals, it could be reasonable to assume that the public's response to the live exports campaign arose from the increased, unprecedented access to visual information. Is the catalyst for the outrage provoked by the *Animals Australia* footage then the simple fact that people are offended and disturbed by the violent abuse of animals? If this were the case, one could surmise that the debate would be motivated by several interests. For example, if there was a generalizable concern over poor animal welfare outcomes, the concern would be directed at *all* animals facing death within these institutions, including the institutions themselves. Correspondingly, the concern would address slaughterhouse practice, regulation and law, *outside* of as well as *within* Australia. The nature of the debate would thus focus on the standard cruelties any slaughterhouse animal faces, not merely those suffered by exported animals. However, the debate never broadened to consider slaughterhouse practices beyond export destinations. Instead, what could certainly be called a moral panic was localized around the treatment of Australian owned animals exported overseas, not a wider concern over the treatment and margin for error within abattoir practice more generally. To underline this, we need only to examine the veritable absence of public and political reaction to allegations of cruelty towards animals *within* Australia during the period in which animal abuse was being exposed in export destinations.

For example, in November 2011, *Animals Australia* publicized footage of pigs being stabbed in the eyes and ears with stunning equipment (March 2003). The same piece of footage revealed another animal being beaten to death with a sledgehammer. In February 2012 *Animal Liberation* uncovered abuse in a Sydney abattoir, making publicly available footage depicting "sheep being hung up and skinned while apparently still conscious, and a man repeatedly belting live pigs over the head with a metal bar" (Rosenberg and Cubby 2012). The footage was the result of an anonymous call from an employee, where the whistleblower warned, "if you thought the footage from Indonesia was graphic, you should see what is happening where I work." In March 2013 *Animal Liberation* publicly aired footage from an Inghams' facility in Tahmoor, one of Australia's largest chicken and turkey meat suppliers, revealing scenes of turkeys being kicked and stomped on, and other turkeys having their feet removed while still alive (Levy 2013). The response by the New South Wales Food Authority distinctly played down any suggestion that these practices were routine or in any way representative of the industry. Peter Day, Director of Compliance, Investigation and Enforcement, stated, "At this point it appears the particular situation relates to individuals and not a failure of systems in place to protect animal welfare" (NSW Food Authority 2012).

Some may argue that these incidents are isolated, and that they do not suggest a systemic norm within Australian slaughterhouses. However, there is ample evidence of systemic welfare violations in Australian slaughterhouses, sometimes verified openly by official agencies. We could refer here, for instance, to the results of the 2012 NSW Food Authority investigations into red meat slaughterhouses in the state. The review uncovered animal welfare breaches in all 10 red meat slaughterhouses in NSW, including the use of ineffective stunning techniques and poor training of staff (Burke 2012). This incontrovertible government finding of what can only be understood as a systemic set of problems in NSW slaughterhouses again failed to generate the same level of government and community concern that was witnessed in response to live exports.

We would suggest that examples of documented cruelty within domestic slaughterhouses, including cruelty supported by similar video footage as that accompanying incidents reported in export destinations, indicate that the success of the live exports campaign cannot be attributed solely to the availability of images that allowed the public to see inside the slaughterhouse. It would thus seem that getting the subject of animal abuse into the cultural, political and legal national psyche is beyond a simple matter of visibility.[2] Why is it, we might ask, that public outrage was not generated to the same extent in the cases of abuse 'at home'?

Responsibility for 'Australian' Cattle

One way in which to consider this atypical reaction to animal abuse is that Australians may feel directly responsible for the cruelty that occurred, as exports are managed by the Australian government's own trade regulations.

---

[2] In this sense we would contest the view that increased visibility necessarily leads to increased protection, as put forward by scholars such as Siobhan O'Sullivan; see *Animals, Equality and Democracy*. In this conversation, Timothy Pachirat offers a useful discussion of visibility and animal welfare in *Every Twelve Seconds: Industrialized Slaughter and the Politics of Sight*.

Otherwise put, Australia is morally liable for what happens to animals abroad as they are in direct control of their export property and what fate these animals meet once they reach destination countries. Australians may consequently feel that the treatment of export animals is within their jurisdiction; that they have the power to halt or prolong the abuse. The public could indeed feel a kind of collective culpability that animals within their care were abused offshore by welfare standards they believe do not meet their own (even if, as we propose above, this notion of national differences in treatment and slaughter practices requires careful and critical questioning). Certainly, Australia exerts substantial capacity to control the conditions of export and slaughter of animals it sends offshore. The government response, particularly its capacity to regulate and cease export, only emphasizes the purchase of a public response to provoke a national public policy reaction. Given the Australian government and other related institutional bodies were (as it was publicly exposed) fully aware of the routine abuses that were occurring, it is striking that the outrage was overwhelmingly directed against the ethics and welfare standards of destination countries. As Siobhan O' Sullivan has frankly put it, "Aren't all these welfare problems a result of the very fact that Australian animals are being exported live, half way around the world?"[3] (2012).

A recurring theme throughout both the animal welfare campaign and the public response was the specific use of the prefaces 'our' and 'Australian' before 'animals' or 'cattle,' reflected in animal advocacy protest literature, placards, in pro-export campaigning, and in the press. The assiduous persistence of these prefaces, and their ubiquity in both anti- and pro- export campaign efforts, is curious. Tellingly, animals abused in Australian institutions do not get so named in the media. The omnipresence and repetition of this discursive trend reveals a clear relevance, but to whom, and for what purpose? Does the preface ('Australian') signal a metonymic slide between 'Australian animal' and Australia? As John Berger has notably detailed, animals typically serve as metaphors, repositories of cultural values, and are often powerfully symbolic of identities; including national identities (1980). Is the origin of the indignation the fact that representatives, ambassadors even, of Australia are being

---

3 Analogously, while a considerable part of the live export campaign focuses on the live export (the shipping of animals over long distances, an additional cruelty on top of that facing them at port) the voyage itself and its myriad animal welfare concerns were not what made headlines and received public concern, the slaughter practices were. A recent example involved the death of some 4000 sheep on a sea vessel. See Narelle Towie, 'More than 4000 Sheep Perish on a Live Export Ship.' In this case the reporting overtly focused on the Jordanian ownership of the vessel, rather than Australian responsibility for the animals.

ill-treated? Culturally positioned as appendage, receptacle or prosthesis, could the abuse of Australian animals be read as an affront to the import and position of Australia more generally, all the more malevolent as it occurred on the international stage? Animal rights campaigns have, however, publicized the arrant cruelty of the kangaroo industry explicitly through utilizing the significance of the kangaroo as a national symbol, yet these interventions have not proved comparably successful in improving welfare outcomes (Burgess 2013). An iconic emblem of national identity, present on the coat of arms and emblazoning Australian currency, the slaughter of kangaroos stimulated a markedly quieter outcry than the killing of sheep and cattle (both of whom had been imported to Australia during colonization by the British, and both 'food' animals, a notoriously difficult category to mobilize support for).[4] As Adrian Franklin has discussed, within Australian culture, animals "do not represent homogeneity but a rather puzzling and unstable heterogeneity" (2006, 14).

Could this adjective, 'Australian', simply dictate ownership: the property right Australians have over animals within their territory? If so, what is its weight as a marker of jurisdiction? And why, then, are Australians not similarly concerned about the mistreatment of the very same property—i.e., 'food' animals—when abuse occurs on Australian soil? In order to think through the property relation between humans and animals, we might consider the complexity of property in its relation to race, as explored by Cheryl I. Harris in her eminent *Harvard Law Review* essay, "Whiteness as Property" (1993). Tracing the history of racialized slavery and the correlation between legal enfranchisement and property rights, Harris observes that the property right of whiteness prevented its owner from being enslaved: "white identity and whiteness were sources of privilege and protection; their absence meant being the object of property" (1721). Harris further observes that like other forms of property, whiteness relies upon a continued expectation of value and value protection, is protected by a right to use and enjoy its privileges, is directly linked to status and reputation which are elsewhere treated as property by defamation law,

---

4   The kangaroo industry in Australia is responsible for the largest commercial slaughter of land-based wildlife on the planet. Kangaroos are hunted for their meat and skin, killed in recreational hunting, and massacred in large numbers when deemed pests. Almost 90 million kangaroos and wallabies have been lawfully killed for commercial purposes in the last 20 years. For more information, see the following reports: THINKK report, 'The Ends and Means of the Commercial Kangaroo Industry: An Ecological, Legal, and Comparative Analysis,' at: https://s3.amazonaws.com/thinkk_production/resources/24/THINK_Kangaroo_Welfare_Report_December_2011_Final.pdf and 'A Shot in the Dark,' by Dror Ben-Ami, at: http://animal-lib.org.au/images/stories/news/a_shot_in_the_dark.pdf.

and like other property, relies upon a right to exclude non-property owners from enjoyment. Harris is not intimating that 'whiteness' is a 'natural' or essential property of its owner. Rather, she is interested in how law and institutions manufacture and fabricate race *as* a property relationship, and the material outcomes this has for bodies and architectures of violence.

Following Harris, one reading we may offer here is that the live exports campaign, and the flurry of regulatory, legal and institutional action which resulted, have aimed to sculpt whiteness, or at best, 'Australian-ness,' as a property relation over animals exported for slaughter. This property relation, which, like 'whiteness,' is not alienable, maintains ongoing property obligations irrespective of who the animal is sold to. These animals are marked (or literally tagged) as exceptional, and thus granted unique status. This property relation works to enable a continuing Australian concern for animal welfare abroad, whilst there is simultaneous immunity conferred to Australia's own welfare practices. Export animals therefore become synonymous with Australian identity as objects of racialized property, with forms of exclusory rights and exemption attached to the indulgence of this property right by the owners. In this sense, we can perhaps more accurately perceive the live exports campaign as not concerned solely with the suffering of animals, but more precisely as fulfilling a metonymic function whose concern is the elaboration of Australian subjectivity and geopolitics.

The success of the manufacturing of live export animals as 'Australian' is not simply a cultural and symbolic fabrication, but a material fabrication articulated through law, regulation, surveillance and tracking. In answer to the 2011 footage, the Federal Department of Agriculture launched an investigation into whether the cattle in the video were in fact 'Australian,' and attempted to trace their journeys through identification checks (SMH 2011). Animals coming from Australia were tagged within abattoirs to differentiate them from locally bred animals and those from other export countries. This was done in order to change and regulate the methods of handling and slaughter; however not for all animals, only those tagged as 'Australian.'

White People Saving White Animals from Non-White People

We can further extend this reading by examining the racialized nature of Australian attempts to liberate export animals, and the ways in which this exercise constructs particular subjectivities and systems of truth. In her canonical piece, "Can the Subaltern Speak?," Gayatri Spivak explores the idea of *epistemic violence* through the example of widow sacrifice in India—the practice of *sati* in Indian communities where a widow would immolate herself on her husband's funeral pyre—and the role of British colonial law in forbidding this practice (1988). Epistemic violence elucidates the ways in which an order of truth allows some subject positions to become intelligible, while obfuscating the possibility of realizing others. In the case of the outlawing of widow sacrifice in India, Spivak demonstrates how this gesture rested upon the reproduction of a truth of British colonizers as saviors of women, and Indians who defended this practice as representing a primitive tradition that supported the killing of women. Spivak elaborates:

> The abolition of this rite by the British has generally been understood as a case of "White men saving brown women from brown men"... Against this is the Indian nativist argument, a parody of the nostalgia for lost origins, "The women actually wanted to die" (93).

In the midst of this interplay that would provide an ideological justification for colonization, the possibility of an account of this experience authored *by women* is completely submerged; a possibility that might have confirmed both that brown women did not want to be saved by British colonizers, nor did they want to die in the context of widow sacrifice. Spivak discerns that "one never encounters the testimony of women's voice consciousness" and that this missing voice may have provided the "the ingredients for producing a countersentence" (93). (That is, a counter sentence to the two competing options: "White men saving brown women from brown men" and "The women actually wanted to die.")

We might playfully amend Spivak's phrase in the context of the live exports discussion to: "White people saving white animals from non-white people." This seems appropriate to describe a phenomenon which is by no means limited to this case study, however palpable in the outpouring of public concern over animals in non-Western abattoirs, animals that Australia seeks to maintain property rights over. Again, we employ 'white' here in the same sense in which Harris does: as a property relation sculpted, reproduced and securitized by law and institutions. Social relations, regulation and institutional power create animals in export destinations as 'Australian' animals that are worthy of saving. At the same time, subjectivities and truths are generated. Firstly, for the concerned Australian who is defined by their desire to save exported animals subjected to particular forms of violent death that are understood as failing to meet Australian standards, and secondly, for the subjectivity of those receiving Australian animals who are characterized as cruel and indifferent to animal suffering.

The media reporting only evinced this rehearsal of Orientalized discourse, whereby export countries lacked the 'civility' and receptiveness to animal suffering that Australians apparently possessed. Some media discussion inferred this as a result of the relative economic poverty of export destinations and an attendant lack of public concern for welfare. However, there have been numerous suggestions that export destinations are 'barbaric' and that wanton cruelty is an inherent trait amongst those receiving 'Australian' animals.

This was illustrated forcefully by Australian Senator Bill Heffernan who insinuated that the only response to allegations of cruelty at export destinations was to 'civilize' the cultures of those who receive Australian animals, a comment which impeccably demonstrates the connotative nature of this debate:

> We're dealing with countries that shoot women when they go to school... Do we end the wheat trade to a lot of the countries in North Africa because they treat their women abysmally and we just turn a blind eye and say it's just a cultural thing?... You've got to learn to bring them along with you, to modernise their abattoirs. (Marx 2013)

What becomes starkly transparent in this reasoning is that concern for forms of cruelty experienced by animals in export destinations are molded by epistemologies which fashion 'civilized' and 'non-civilized' actors, and thereby create the possibility for the alleviation of animal cruelty to be read as a redemptive act by those same civilized actors. The sentence "White people saving white animals from non-white people" establishes a truth that Australians kill "their own" animals in a civilized way, while non-white others do not. This prescription thus determines for us that instances of inhumane killing within Australia are treated as anomalous and exceptional, and routine practices in the context of animal slaughter, such as stunning, are rendered "humane" and impervious to critique (see Probyn-Rapsey, 2013). Simultaneously, examples of inhumane treatment in export destinations become symptomatic of systemic practice. In either case the truth that 'we' kill 'our' animals in a 'humane way' is confirmed. It is unsurprising that commentators metronomically repeat sentiments that only validate that the live exports campaign conforms to the kind of understanding we have laid out above. Considering slaughter practices in export destinations and the success of the campaign, the words of one commentator, Clive Phillips (2011), seamlessly embody this overtone, "Slaughter overseas had all the right ingredients to raise concerns in the community: painful procedures routinely conducted on defenceless cattle by foreigners."

'Animal Nationalism'

Of course, the live exports debate both arises from and gets interpolated through a pre-existing context of a global geopolitics of animal slaughter and racialized subjectivities. Australia is a meat exporting powerhouse: alongside Brazil, India and the United States, Australia is one of the largest exporters of beef in the world (Beef Central 2012). In response to a persistent increase in global per capita meat consumption, particularly concentrated in the Asia Pacific region, Australia is continually establishing itself as a source that can meet international demands. Recent moves to generate Australian free trade agreements with Japan and China have included a focus on creating export markets for Australian meat and dairy (Tehan 2014; Kenny and Wen 2014). Synchronously, global migration flows are being supported to secure the labor force to expand animal production and slaughter. For instance, Australia recently created meat industry labor agreements generating a temporary migrant labor scheme for skilled laborers to work in Australian abattoirs (Department of Immigration and Border Protection 2013). Live export regulation is part of this story. Given Australia's global positioning, the imperative for industry and government will be to secure supply chains and minimize public opposition to live export. Not only are producers and governments keen to ensure a secure supply of live animals to export countries that have the capacity and value chains to efficiently slaughter animals themselves, Australia is overtly seeking to use its supply chain regulation over 'Australian' animals to influence slaughter and welfare practices in receiving destinations. In a perverse sense, this civilizing mission has come into being largely in response to calls from live export critics for a wide scale ban. Against this view, it has been argued that welfare will only be achieved if animals from Australia continue to be exported and slaughtered overseas (Coombes 2012). As one critic has suggested:

> The absence of Australian livestock lessens Australia's ability to legitimately work to improve facilities, education, training and monitoring. Removing Australian governmental, bureaucratic and diplomatic involvement in foreign markets removes our formal ability to encourage and implement change. (Flint 2013)

Industry and government have similarly made claims that their goal is to use the live export industry to improve conditions of slaughter at export destinations. The effects of this move can be seen in industry training programs for workers in these countries. The Australian Live Exporters Association has even claimed, "We are the only country that exports cattle overseas and also puts an investment into ensuring animal welfare in the market... In the last two years 3000 staff in Indonesia have been trained up by us" (Black 2013). Government has also responded: in 2013 Australia had allocated some 10.2 million dollars in funds in foreign aid spending to assist eligible countries "to... apply the World Organisation for Animal Health's animal welfare guidelines on livestock transport, slaughter of animals for human consumption and handling at feedlots" (Aston 2013). Here, the enhancement of 'animal welfare' standards in export destinations, driven by Australian public interests in preventing animals experiencing cruelty at the hands of 'foreigners,' is openly deployed as a tool to leverage the expansion of global industrialized meat production.

Jasbir K. Puar has conceptualized "homonationalism" as a method to realize how a gay and lesbian rights agenda is put into service by the nation State (2007; 2012; 2013). This is enacted in order to situate and generate ostensibly tolerant and accepting and non-tolerant and non-accepting cultures, whilst building solidarity across borders to further hegemonic interests. Puar is interested, for example, in why it is that U.S. foreign policy and its advocates might focus on gay and lesbian rights in Iran or the occupied territories, rather than elsewhere (including the U.S.), and how this focus invents a dichotomy between 'gay friendly' and 'non-gay friendly' countries. Homonationalism, she argues, "has come to structure the conditions of possibility about sexuality and rights internationally," as we inhabit a historical moment where "the right to, or quality of sovereignty is now evaluated by how a nation treats its homosexuals" (2012). Similar to the many critiques of Western feminist gestures to 'rescue' Middle Eastern women, Puar outlines how the (appearance of a) tolerance of queerness is now a necessity in ratings of a culture's democratic and moral civility. Invariably, these distinctions are Orientalist in nature. They produce an imaginary of 'civilized' accepting cultures juxtaposed against 'barbaric' cultures, and spawn a number of effects, including 'justified' demands for civilizing responses. In a seemingly inexhaustible reiteration of this same logic, here it is animal welfare (or potentially, animal rights) that is positioned within a global geopolitics to generate racialized and nationalistic boundaries, creating the epistemic framing of civilized versus barbaric, advanced versus backwards, and therefore, those lives deemed either worthy or unworthy of being saved. The morbid sardonicism this script decrees is that the animal (or woman, or sexual other) that is 'saved' by this discourse is patronized and made diminutive, a victim reliant upon others for her salvation. We might usefully adopt the term 'animal nationalism,' as used by scholars such as Janet Davis (2013) and Colin Salter (2015), to describe this racialized discourse surrounding animal treatment in its relationship to culture.

Akin to Puar's critique of 'pinkwashing,' certain practices of consumer-driven animal welfare in Australia fit neatly into a larger discourse of 'greenwashing' (2012). A range of methods, advertising motifs and promises that companies or institutional bodies make in order to appear more environmentally friendly, greenwashing instead misleads the public by feigning an ethical response and offering it as a genuine alternative to complicity. Operating in tandem with this trend, there is a recent burgeoning market, especially in consumer goods, aimed to capitalize on a demographic that wishes to consume products deemed 'animal friendly.' Predictably, most of the avenues provided to the public as techniques to reduce animal suffering are opportunities for 'value added' product differentiation and niche marketing for profit (such as paying more for 'free range' eggs) and do little to fundamentally challenge structural animal abuse and our wide scale support of it. Placing a higher price tag on products deemed 'animal friendly' compels us to choose between civility and barbarism (as if this division were self-evident), and to do so with our dollar, sending out a very clear message: one should be willing to pay for their ethics. This sleight of hand makes it appear as if the most ethical and charitable citizens are naturally the most wealthy, and, in so far as 'civility' is racialized, these consumers might be rendered as 'white.'

The desire to consume 'animal friendly' products, or to "humanely kill animals at home," or even to seek 'ethical meat,' relies on a certain idealized view of one's culture as enlightened, caring and ethical (Stănescu 2011). What is noteworthy about the aspiration to forge such a national self-image is that it requires an opposite against which it is to be defined. This divide enables the realization of certain objectives: assuring Australia's place as a modern, principled culture, discursively producing Middle Eastern and Asian export nations as savage and regressive, and thus reaffirming a cultural gulf between the two. That is, Australia's moral superiority emerges through, and is affirmed by, its creation and public shunning of the uncivilized Oriental other. What simultaneously transpires is a narrative of naturalization which implicitly dictates that Australians are naturally more compassionate, more properly developed; more human, as it were.

The abuse of animals overseas produces a racialized subject, with an explicit reference point one can nod to in such a way that the footage Australians witnessed acts as evidence to authorize what was an ideological position as an apolitical factual description: "they *are* cruel, they *are* barbaric." Presenting non-white cultures as in need of moral redemption from Australia also helps Australians to better digest animal abuse themselves, diluting the emphasis from their own implicatedness and eliding their own national track record, punctuated as it is with evidence of clear and ongoing cruelty.

As aforementioned, what does not get called into question throughout this conversation over who kills 'their' animals the 'right way' is the fundamental question of why it is that humans have an unquestioned entitlement to kill animals, an entitlement that is (even if cursorily at times) deemed abhorrent when directed toward humans. Rather, what we are left with is the tacit acceptance of the fact that animals do not require saving in general, they require saving from non-white others. As with Spivak's widowed women, then, the way this discourse carves out its path never seriously considers, in any substantial way, the plight and the lives of animals, as whether animals are killed overseas or in Australia, the sovereign right *to* kill remains undisturbed.

Postscript

After the federal budget speech in 2013, the then shadow treasurer, Joe Hockey, was interviewed by the Australian broadcaster Alan Jones. In the interview, Jones referred to a crisis meeting of cattle farmers in Queensland affected by the live export bans. The attendees highlighted a collapse in incomes and an inability to feed their cattle. At the same time, Jones pointed out that the federal government had proposed raiding the foreign aid budget to meet the costs of accommodation, food, clothing and other basic necessities for asylum seekers. Linking these two seemingly unconnected aspects of public policy, asylum seekers and live exports, Jones asked of Hockey, "Why can we feed asylum seekers and not farmers and cattle?" (2013). Surely, this question exemplifies a kind of circular completion of this game of truth, where the racialized politics of asylum seekers comes head to head with the racialized politics of live export. Animal advocates are traditionally accustomed to confronting an unshakeable conviction from governments, industry and the public, which unhesitatingly demands that in matters of ethics, humans precede animals. This conviction poses a challenge for those interested in addressing violence towards animals, since a human investment in the continuation of slaughter, experimentation or recreational use of animals will unswervingly trump consideration of animal suffering. However here, in Alan Jones' plea—"why can we feed asylum seekers and not farmers and cattle?"—we find a perfect reversal of the order of things achieved through recourse to racialization, where 'our' cattle may prove to have lives more worthy than some humans. This question—"why can we feed asylum seekers and not farmers and cattle?"—represents, it would seem, a new inflection of epistemic violence in this discussion, as it manages to perfectly silence not only animals, but racialized human minorities who find themselves quite arbitrarily conjoined as speechless within this discourse.

In line with the vast shifts in production and supply chains accompanying globalizations, will the production of meat increasingly move to low wage cost and low welfare regulation environments in order to increase profitability, and more conspicuously, to sidestep the scrutiny of Western animal activism? If so, will we see a parallel market arise where consumers will pay more for animal products produced in the West as a guarantor of better animal treatment? Alternatively, will interventionist exporters work with export destinations to create 'value added' purportedly 'high welfare' operations elsewhere in order to cement mass scale transport of live animals? What forms these narratives may take in the future is worth pondering. Regardless, the more seamlessly animal welfare is coded as a marker of modernity, the more violently it can be mobilized upon the racialized politics of the world stage and enlisted towards the expansion of the global meat industry, and the intricate value chains that support its profitability.

References

Animals Australia and RSPCA. "Live Exports to Indonesia." 2011. http://banliveexport.com/documents/FactSheet-cases.pdf. Accessed June 9, 2015.

Aston, Heath. "Foreign aid for climate cut as live exports get funding." *The Age*. May 18, 2013. http://www.theage.com.au/federal-politics/political-news/foreign-aid-for-climate-cut-as-live-exports-get-funding-20130517-2js6o.html. Accessed June 9, 2015.

Beef Central. "India Takes Over as World's Largest Exporter of Beef." *Beef Central*. October 4, 2012. http://www.beefcentral.com/trade/india-takes-over-as-worlds-largest-beef-exporter/. Accessed June 9, 2015.

Berger, John. "Why Look at Animals?" In *The Animals Reader: The Essential Classic and Contemporary Writings*, edited by L. Kaloff and A. Fitzgerald, 251–261. Oxford: Berg, (1980), 2007.

Handle With Care Coalition. *Beyond Cruelty, Beyond Reason: Long Distance Transport and Welfare of Farm Animals*. London: World Society for the Protection of Animals: 2008.

Black, Simon. "Animal Justice Party Pledges to Shut Down Live Exports with Industry Campaign." *The Daily Telegraph*. September 4, 2013. http://www.dailytelegraph.com.au/news/nsw/animal-justice-party-pledges-to-shut-down-live-exports-with-industry-campaign/story-fniocx12-1226710074864. Accessed June 9, 2015.

Burgess, Elise. "National symbol but treated cruelly at home." *Voiceless*. April 8, 2013. https://www.voiceless.org.au/content/national-symbol-treated-cruelly-home. Accessed June 9, 2015.

Burke, Kelly. "Multiple Deficiencies Uncovered in NSW Abattoirs." *The Sydney Morning Herald*. May 18, 2012. http://www.smh.com.au/environment/animals/multiple-deficiencies-uncovered-in-nsw-abattoirs-20120517–1ytn0.html.

Coombes, Stephanie. "Live Export Promotes Animal Welfare." *The Daily Telegraph*. November 7, 2012. http://www.dailytelegraph.com.au/news/opinion/live-export-promotes-animal-welfare/story-e6frezzo–1226511677639. Accessed June 9, 2015.

Davis, Janet M. "Cockfight Nationalism: Blood Sport and the Moral Politics of American Empire and Nation Building." *American Quarterly*. 65.3, 2013.

Deckha, Maneesha. "Toward a Postcolonial Posthumanist Feminist Theory: Centralizing Race and Culture in Feminist Work on Nonhuman Animals." *Hypatia: Journal of Feminist Philosophy* 27.3 (2012): 527–545.

Department of Immigration and Border Protection. *Meat Industry Labour Agreements: Information for Employers about Labour Agreement Submissions*. Australian Government, September 2013.

"Extend cattle tracking, Nats says." *Sydney Morning Herald*. June 15, 2011. http://www.smh.com.au/environment/animals/extend-cattle-tracking-nats-say-20110615-1g2qb.html. Accessed June 9, 2015.

Flint, Nicolle. "ABC Loses its Balance Over Animal Welfare." *The Age*. May 9, 2013. http://www.theage.com.au/comment/abc-loses-its-balance-over-animal-welfare-20130508-2j7va.html. Accessed June 9, 2015.

Franklin, Adrian. *Animal Nation: The True Story of Animals and Australia*. University of New South Wales: University of New South Wales Press, 2006.

Haraway, Donna. *When Species Meet*. Minneapolis: University of Minnesota Press, 2008.

Harper, Amy Breeze. "Race as a 'Feeble Matter' in Veganism: Interrogating Whiteness, Geopolitical Privilege, and Consumption Philosophy of 'Cruelty-Free' Products." *Journal for Critical Animal Studies* 8.3 (2010): 5–27.

Harris, Cheryl I. "Whiteness as Property." *Harvard Law Review* 106.8 (1993): 1707–1791.

Jones, Alan. Alan Jones Interview—Joe Hockey. *2GB Radio*. May 15, 2013. Podcast available at: http://www.2gb.com/audioplayer/9080#.U6LlHfmSyP8. Accessed June 9, 2015.

Kenny, Mark, and Wen, Phillip. "Tony Abbott, Chinese Eager to Finalise Free-Trade Deal." *Sydney Morning Herald*. April 9, 2014. http://www.smh.com.au/federal-politics/political-news/tony-abbott-chinese-eager-to-finalise-freetrade-deal-20140409-zqsqh.html#ixzz355Z7K77w. Accessed June 9, 2015.

Kim, Claire Jean. *Dangerous Crossings: Race, Species, and Nature in a Multicultural Age*. Cambridge University Press, 2015.

Kim, Claire Jean. "Multiculturalism Goes Imperial: Immigrants, Animals, and the Suppression of Moral Dialogue." *Du Bois Review* 4.1 (2007): 1–17.

Kim, Claire Jean. "Slaying the Beast: Reflections on Race, Culture and Species." *Kalfou: A Journal of Comparative and Relational Ethnic Studies* 1.1 (2010): 57–74.

Kymlicka, Will, and Donaldson, Sue. "Animal Rights Multiculturalism and the Left." *Journal of Social Philosophy*. 45.1 (2014): 116–35.

Levy, Megan. "'Torture for Fun': Police Given Shocking Abattoir Footage." *Sydney Morning Herald*. March 21, 2013. http://www.smh.com.au/national/torture-for-fun-police-given-shocking-abattoir-footage-20130321-2ggm2.html#ixzz3556CdqBV. Accessed June 9, 2015.

March, Stephanie. "Abbatoir Closed Over Animal Cruelty Concerns." *ABC News Online*. Nov 30, 2011, http://www.abc.net.au/news/2011-11-30/abattoir-closed-over-animal-cruelty-concerns/3703144. Accessed June 9, 2015.

Marx, Jack. "Senator Blames Barbaric Culture for Cattle Cruelty." *ABC News Online*. May 8, 2013. http://www.abc.net.au/news/2013-05-08/senator-blames-barbaric-culture-for-cattle-cruelty/4676704. Accessed June 9, 2015.

New South Wales Food Authority. NSW Food Authority Response to Animal Cruelty Allegations at an Inghams Tahmoor Processing Facility. *NSW Food Authority*. 21 March 2012. http://www.foodauthority.nsw.gov.au/news/media-releases/mr-21-Mar-13-animal-cruelty-allegations-inghams/#.UmhqGPlgeiA. Accessed June 9, 2015.

O'Sullivan, Siobhan. *Animals, Equality and Democracy*. Palgrave Macmillan, 2011.

O'Sullivan, Siobhan. "Live Animal Exports and Australian Politics—More than a Case of Conscience." *The Conversation*. August 23, 2011. http://theconversation.com/live-animal-exports-and-australian-politics-more-than-a-case-of-conscience-2964. Accessed June 9, 2015.

O'Sullivan, Siobhan. "The Live Export of Animals Will Always be a Bloody Business." *The Conversation*. November 6, 2012. http://theconversation.com/the-live-export-of-animals-will-always-be-a-bloody-business-10547. Accessed June 9, 2015.

Pachirat, Timothy. *Every Twelve Seconds: Industrialized Slaughter and the Politics of Sight*. New Haven: Yale University Press, 2011.

Phillips, Clive. "Why We Need an Independent view of Live Animal Exports." *The Conversation*. June 23, 2011. http://theconversation.com/why-we-need-an-independent-view-of-live-animal-exports-1960. Accessed June 9, 2015.

Probyn-Rapsey, Fiona. "Stunning Australia." *Humanimalia*. 4:2, 2013. 84–100.

Puar, Jasbir K. "Rethinking Homonationalism." *International Journal of Middle East Studies* 45 (2013): 336–9.

Puar, Jasbir K. *Terrorist Assemblages: Homonationalism in Queer Times*. Durham: Duke University Press, 2007.

Puar, Jasbir K., and Mikdashi, Maya. "Pinkwatching and Pinkwashing: Interpenetration and Its Discontents." *Jadaliyya* Aug 09, 2012. http://www.jadaliyya.com/pages/index/6774/pinkwatching-and-pinkwashing. Accessed June 9, 2015.

Puddy, Rebecca. "Ritual Animal Killing Routine in Australia." *The Australian*. June 23, 2011. http://www.theaustralian.com.au/national-affairs/ritual-animal-killing-routine-in-australia/story-fn59niix-1226080205383. Accessed June 9, 2015.

Rosenberg, Jen, and Cubby, Ben. "Covert Evidence of Cruelty Halts Abbatoir." *Sydney Morning Herald*. February 10, 2012. http://www.smh.com.au/environment/animals/covert-evidence-of-cruelty-halts-abattoir-20120209-1rx7w.html. Accessed June 9, 2015.

Salter, Colin. "'Our' Whales and Cows: Identity and Nationalism in Australian Moral Outrage." Public Lecture. Group for Society and Animal Studies. Universität Hamburg. 20th October 2015.

Spivak, Gayatri Chakravorty. "Can the Subaltern Speak?" In *Marxism and the Interpretation of Culture*, edited by C. Nelson and L. Grossberg, 271–313. Basingstoke: Macmillan Education, 1988.

Stănescu, Vasile. "'Green Eggs' and Ham? The Myth of of Sustainable Meat and the Danger of the Local." In *Critical Theory and Animal Liberation*, edited by John Sanbonmatsu, 239–56. Plymouth: Rowman and Littlefield, 2011.

"Sydney Protestors Call for Live Exports Ban." *Sydney Morning Herald*. April 14, 2011. http://www.smh.com.au/environment/animals/sydney-protesters-call-for-live-export-ban-20110814-1isrz.html. Accessed June 9, 2015.

Tehan, Dan. "Japan Free Trade Agreement Giving Long-Term Hope to Australian Farmers." *Weekly Times Now*. April 21, 2014. http://www.weeklytimesnow.com.au/news/opinion/japan-free-trade-agreement-giving-longterm-hope-to-australian-farmers/story-fnkerdb0-1226891085449. Accessed June 9, 2015.

Towie, Narelle. "More than 4000 Sheep Perish on a Live Export Ship." *WAToday*. January 16, 2014. http://www.smh.com.au/environment/animals/more-than-4000-sheep-perish-on-live-export-ship-20140116-30wf2. , 2015.html

RSPCA. "What is Halal Slaughter in Australia?" RSPCA Australia Knowledgebase. http://kb.rspca.org.au/What-is-halal-slaughter-in-Australia_116.html. Accessed June 9, 2015.

Wadiwel, Dinesh J. *The War Against Animals*. Leiden, The Netherlands: Koninklijke Brill, 2015.

CHAPTER 5

The Whopper Virgins: Hamburgers, Gender, and Xenophobia in Burger King's Hamburger Advertising

Vasile Stănescu

In 2008, Burger King began a new advertisement campaign entitled *The Whopper Virgins*, which purported to go to the "the most remote parts of the world" to discover people who "did not even have a word for hamburger" (Crispin, Porter & Bogusky 2008). The purpose for these travels was so that Burger King could conduct the "purest taste test in the world" (Ibid). The ads were filmed in Thailand, Greenland, and Romania. This ad campaign was one of the most successful in Burger King's history, receiving multiple awards, significant web traffic, widespread media attention, and correlating with one of the largest stock price increase in the company's history (Trosclair 2009; York 2009).

FIGURE 5.1   *Burger King's* Whopper Virgins *advertising campaign.*

In this chapter, I argue that the success behind this campaign was not tied to a taste test as the ads claimed; instead, the advertisements were effective because of the linkages they made between the consuming of meat from Western style fast food restaurants and the stereotype of 'the effeminate rice eater' which has a long history of being deployed as a rhetorical means to naturalize colonialism and xenophobia. This chapter focuses specifically on the ads of 'Transylvanian farmers' that were filmed in my family's homeland of Maramureş, Romania. My argument is that such a campaign worked because the stereotypes between meat eating, gender, and xenophobia continue to resonate with a broad section of the public in the United States. These ads mask the hunger and poverty of the so-called 'third world' in a naturalized discourse based on supposed deficiencies in diet. Consequently, *The Whopper Virgins* refigure the intrusion of the Western style fast food diet from a neoliberal imposition into an act of 'humanitarian' assistance.

The Purest Taste Test in the World

The Whopper Virgins began in 2008 and consisted of two parts: first, a series of television commercials each beginning with the claim, "We traveled to the most remote parts of the world, to people who did not even have a word for hamburger, and we asked them to compare the Whopper and the Big Mac for the purest taste test in the world!"; and second, a 'documentary' produced by Burger King, which revealed the results of the 'experiment.' Each of the nearly identical ads started with images of the peoples of Thailand, Greenland, and Romania shown in traditional clothing and living an agrarian lifestyle removed from technology. Burger King then flew selected individuals to a modern boardroom where the taste testers offered them two burgers (a Whopper and a Big Mac). Many of the people were shown not being able to figure out how to eat a hamburger. The ads informed the viewers that the majority chose the Whopper. In the second half of the documentary, the film crew flew back into the 'villages' of the participants who were provided with free Whoppers.

A God-Given Right to a Juicy Hamburger

The Whopper Virgins represented an influential, highly effective, and award winning advertising campaign with deep and emotional resonance to large sections of the American public. Nevertheless, there were serious concerns voiced; for example, the *New York Daily News* interviewed Sharon Akabas, the associate director of the Institute of Human Nutrition at Columbia University about her outspoken criticisms of the advertisement (York 2008). She claimed in this interview that she received thousands of letters directed against her by consumers in the United States because of her criticism of the ads. One letter attacked Akabas for depriving other people in the world of their "God-given right to a juicy hamburger."

Statistics also demonstrate that Burger King's *The Whopper Virgins* represented a uniquely successful advertising campaign. Between December 3 and December 31, the month of the ad campaign, Burger King's stock rose from $18.78 to $22.82, an increase of 21.5% (Trosclair 2009)—one of the largest stock increases in the history of Burger King. (To provide a comparison, the Dow Jones Industrial average rose only 2.1% in that same period.) Likewise, during a single month, the Burger King ad campaign generated over a million views of the online documentary (York 2009). Based on these results, the advertising campaign garnered multiple top awards for effective advertising, including the Cannes, the Webbey, and London Advertising Award (Caiozzo 2012). The following message appeared on the website for Paul Caiozzo, Associate Creative Director of Crispin, Porter and Bogusky, the ad agency who created *The Whopper Virgins*:

> The highlights from Paul's three years at Crispin include... *Whopper Virgins*, a controversial campaign discussed on CNN, praised by *National Geographic*, and parodied on *Saturday Night Live*. After this run, Burger King was named marketer of the year, Crispin was named Agency of the Year, and Paul was named one of the top creatives in the world.

The point is that these ads *worked*. While controversial, they appealed to people: they garnered media attention, won awards, and increased both stock prices and revenue for Burger King.

Transylvania Farmers

> We are in Transylvania; and Transylvania is not England. Our ways are not your ways, and there shall be to you many strange things.
> BRAM STOKER'S *Dracula* 2012, 31

∴

The penultimate episode of the series of ads begins: "What happens if you take Transylvanian farmers who have never eaten a burger and ask them to compare Whopper versus Big Mac in the world's purest taste test?" (Crispin, Porter & Bogusky 2008). (To state perhaps the obvious, there is no country called 'Transylvania'; 'Transylvania' refers to the center forested region of Romania. It has no connection with the traditional myths of 'vampires' or the historical figure of 'Dracula'). The caption of the ad clarifies the name of the town—'Budeşti'—but Romania contains over a dozen villages, towns, and communes known as 'Budeşti,' none of which lie within the 'Transylvanian' region of Romania. Furthermore, the traditional Romanian clothes shown in the film—clothing rarely worn in Romania—suggests that the piece was actually filmed in Maramureş, which has retained much of its original cultural heritage due to its enclosure by the Eastern Carpathians. I particularly suspect this is the case because there is a commune in the county of Maramureş called 'Budeşti,' which is also the site of the wooden church of Saint Nicholas built in 1643. In the wake of the controversy over the ads, Burger King attempted to justify its actions by contending that "[t]he company helped fund restoration of a 17th century church in Romania" (Burger King Corporation 2009).

I take this time in discussing where Burger King would have had to film the piece in order to make the point that the ad's sole concern was to exoticize Romania by splicing together the traditional dress and architecture of Maramureş with the fantastical sound of 'Transylvania' even though Maramureş and Transylvania are in different parts of the country. The fact that the documentary exoticized the location is a clear indication that other claims by the documentary were equally incorrect and exoticized, such as the insinuation that no one in 'Transylvania' owned a television, had ever seen an advertisement for either Burger King or McDonalds, or ever eaten (or seen) a hamburger. In my experience such contentions are false because, thanks to grants from Stanford University and the *Institutul Cultural Român*, I visited and studied in the Maramureş region of Romania from which my family emigrated. And while it is true that Maramureş is a predominately agriculturally-based area, many people own a television, know all about McDonalds and Burger King, and have regularly eaten hamburgers. This would also be the case throughout the 'Transylvanian' region of Romanian, as well as the rest of the country. In fact, McDonalds has been in Romania for over fifteen years and operates over sixty-one restaurants across the country, which have served over 500 million customers (Business Review 2010).[1] (While those numbers may

---

1 See also *Doing Business in Romania*. "McDonald's Romania invests EUR 9m in Q1 to spread tentacles in the field." May 19, 2009 http://www.doingbusiness.ro/en/business-news/11742/mcdonalds-romania-invests-eur-9m-in-q1-to-spread-tentacles-in-the-field. (June 1, 2012).

seem conservative by American standards, it is important to remember that Romania is only about the size of the state of Tennessee). And, McDonalds specifically operates multiple restaurants throughout the regions of both Maramureş and 'Transylvania' (depending on where we wish to believe that the ad was actually filmed).[2] There were also, at the time of the ad campaign, eight Burger King restaurants in Romania. Burger King even released a press release less than a year before *The Whopper Virgin* ad celebrating the fact that they were "increasing the Burger King brand throughout Romania" (Burger King Corporation 2012).[3]

The inaccuracies between the claims in the ads and the reality of Romania would have been obvious to everyone making the ad—the film crew would probably have driven past multiple McDonalds or Burger King outlets, as well as ads for both these companies, in the process of filming the advertisements. However, despite the ensuing controversy, Burger King has refused to admit that their claims concerning Romania were false. For example Russ Klein, the president of global marketing strategy and innovation for Burger King Corporation, defended the Whopper Virgins campaign by stating: "During a time when consumers are craving it most, honesty and transparency are the heart and soul of this campaign" (Burger King Corporation 2009).

Effeminate 'Rice Eaters'

During the 1800's, the 'effeminate rice eater' represented a widespread and well known colonial stereotype based on the argument that it was the eating of meat that helped colonizers to become the more masculine, and therefore, the more dominant, power in the colonial age, versus the supposedly 'effeminate' rice and corn eaters of the recently colonized countries. This trope filled the research of the 19th century and helped to justify colonialism under a scientific ideal based on the supposed failure of non-Western nutrition and particularly the argument that these other people did not consume enough and/or the right type of meat as their Western counterparts. For example, J. Leonard Corning, a well-respected medical researcher and doctor, composed a monograph in 1884 entitled *Brain Exhaustion*, in which he argued that the colonial population lacked the "intellectual vigor" of the English, not for racial reasons,

---

2   See McDonalds, Romania, http://www.mcdonalds.ro/ (June 1, 2012).
3   Translation my own, literal text "Remus Tiucăă, General Manager Atlantic Restaurant System a spus: 'Suntem foarte bucuroşi căă reuşim săă contribuim la creşterea brandului Burger King în Romania."

but because they did not eat the enough of the right types of Western meat. As he wrote in a passage representative of his work as a whole:

> Thus flesh-eating nations have ever been more aggressive than those peoples whose diet is largely or exclusively vegetable. The effeminate rice eaters of India and China have again and again yielded to the superior moral courage of an infinitely smaller number of meat-eating Englishmen... But by far the most wonderful instance of the intellectual vigor of flesh eating men is the unbroken triumph of the Anglo-Saxon race. Reared on an island of comparatively slight extent, these carnivorous men have gone forth and extended their empire throughout the world. (Corning 1884, 196–7)

What is important to understand is that such ideas did not represent discredited or fringe ideas of the scientific establishment; and neither were such ideas understood by their practitioners as explicitly racist or colonialist. Indeed, quite the opposite. The motif of the 'effeminate rice eaters' was instead regarded as an intelligent argument in nineteenth century Europe—a concept that reiterated the biases of colonialism and sexism under a supposedly non-racist and non-colonialist worldview based on the mutable characteristic of diet instead of an immutable genetics (Corning 1884, 196–7). Part of the very *appeal* of the notion of the 'effeminate rice eaters' is that it represents an area of possible change and improvement in contradistinction to the notions of the colonial subject found in treaties on genetic racism that could not be changed. The idea of nutritional deficiency tied to meat seemed to offer the colonialist scientists a 'solution'—colonized populations could be 'helped' if they were simply provided the right kinds and amount of Western style meat. As Rachel Laudan has chronicled in some detail:

> For many scientists, politicians and writers, this democratized power diet of white bread and beef, rather than climate or heredity for example, explained the West's industrial prowess, intellectual achievements, and above all its overseas empires. Sarah Hale [an influential American writer and editor] reminded her readers that the "portion of the human family, who have the means of obtaining [animal] food at least once a day... hold dominion over the earth. Forty thousand of the beef-fed British govern and control ninety million of the rice-eating natives of India."... A well-known Australian doctor assured the readers of his dietary text that "Rice is, from an economical point of view, a wretched article of diet... We might expect to find rice-eaters everywhere a wretched, impotent, and effeminate race, and such is the case." (2001, 4–6)

Moreover, these claims about meat eating, gender, and race operate not only in expansionist forms of colonialism but also internally against immigrant groups within the United States and particularly against the Chinese. Meat eating itself became articulated in the 19th century as an example of 'white privilege' that differentiated American workers from the immigrant Asians who were, again, cast as 'effeminate rice eaters.' In other words, racial and diet stereotypes, the colonial justification of European paternalism, and an internal hostility to immigrants became interwoven into a single worldview which portrays all immigrants as both biologically limited and threatening to white, American manhood, because they did not eat enough of, or the right types of, Western style meat. As E. Melanie DuPuis had documented:

> The working class responded [to immigration and cuts to wages] by defending its right to eat meat, as a privilege of white citizenship. The locus of this conflict became race, specifically [targeting] the Chinese.... White working class men deployed nativist anti-Chinese arguments in their demands for a living wage that would support their meat eating. Rejecting nutritionists' arguments that a meat-heavy diet was bad for them, the representatives of the newly established workers' organizations struck back, on behalf of meat [consumption] and [non-immigrant] working class jobs. (2007, 40)

DuPuis claims that it was not coincidental that the colonialism, nativist union sentiment, and the decrease in the cost of meat occurred simultaneously. Instead, she suggests that they forged a mutually beneficial relationship in which eating meat—a large amount of meat and the right type of meat—became a symbolic proxy for the issues of class, gender, colonialism, and race privilege as they impacted the displaced white, male, middle class worker. Market forces helped to allay working class fears not by improving real wages or conditions for workers but simply by providing them with ever-greater amounts of increasingly cheap meat. As she writes, "Workers did not exactly win the fight over wages, but they did win the fight to eat meat, if not in the way they had imagined. Meat became exceedingly cheap, a political bargain between farmers growing grain and workers desiring the tasty marbling of grain-fed beef" (DuPuis 2007, 43).

'Whitewashing' Xenophobia

It is my assertion that *The Whopper Virgins* advertising campaign exoticized contemporary populations in order to recreate these xenophobic reasons for

FIGURE 5.2   Real Locations. Real Burgers.

eating Western style meat and justify their imposition of fast food restaurants as their own type of pseudo-humanitarianism. In the case of the ads featuring Thailand and Greenland, the linkages to pre-existing discourses on 'rice eaters' seems clear since the prior stereotypes focused primarily on colonialized people of color and, particularly, those of Asian descent. Less immediately clear are the linkages to the 'Transylvanian farmers' from Romania. Indeed, at first glance, Romania would seem to have nothing to do with the original stereotype of the 'effeminate rice eater' in that Romanians are usually coded as 'white' in the West and Romania, as part of Eastern Europe, is not generally considered a colonialized country. However, I suspect that Romania was deliberately chosen for these reasons, as a type of 'white cover' to hide behind. In other words, the ad attempts to immunize itself against charges of 'racism' and 'colonialism' by including stigmatized and mocked white 'exotic' groups (i.e. 'Transylvanian farmers' who have little to do with Transylvania).

Burger King's media creators redeploy the traditional stereotypes of colonialism, racism, and xenophobia, in which people from other cultures were displayed as unindustrialized, exotic, ignorant, and thus, comedic. And yet these oft-repeated tropes are now 'hidden' since some (although not all) of the people being mocked are 'white.' Indeed, the ads deploy the traditional and stereotypical, minstrel, and carnival motifs documented in texts such as Bernth Lindfors' *Africans on Stage* (1999), Jan Nederveen Pieterse's *White on Black* (1998), and the edited collection *Inside the Minstrel Mask* (1996), including claims of education for the non-Western audience; cross culture communication; derogatory humor; and exoticism redeployed to a (at least partly) 'white' group. The change itself is not such a large alteration as the original minstrel shows themselves frequently included derogatory portrays of multiple ethnicities and races, including those who would now be coded as 'white' (Bean, Hatch and McNamara 1996, xiii).

I think we can witness this same technique of 'white cover' most clearly in the movie *Borat*. Like *The Whopper Virgins*, *Borat* was a wide success to Western audiences (Low and Smith 2007, 31). And, as with the *Whopper Virgins*, part of this success seems premised on archaic and stereotypical views of Eastern Europeans as backward, ignorant, uneducated, and, therefore, comedic. For example, Borat is shown as not even knowing how to use a toilet and, therefore, brings down a bag of his own excrement to the dinner host (Baron Cohen et al 2007). While Sacha Barron Cohen's movie is a multilayered critique/exposé of the prejudicial views held by the people 'Borat' interacts with, at the same time, it is not entirely clear if the major humor and appeal stems from laughing *with* Borat as much as a laughing *at* Borat.[4] This supposed ignorance (not being able to figure how to eat a hamburger, not being able to figure out how to use a toilet) is so extreme that it seems a (re)deployment of the minstrel themes, rendered socially appropriate, by their application to now 'white' characters. It is my contention that *The Whopper Virgins* engages in 'white washing' via its minstrel show themes. As such it attempts to hide the (ongoing) stereotypical and comedic exoticization of people of color by also criticizing people coded as 'white' while, at the same time, the 'white' populations of Romania are presented only via the traditional racist characters of the comedic minstrel show. The United States has a very long history of showing the conquered subject from wars in a demeaning manner for the public's enjoyment.[5] Following the United States 'victory' of the Cold War, media products such as *Borat* and *The Whopper Virgins* seem to trade in a similar intentional demeaning of the 'conquered subject' of the, now fallen, Eastern European who can be artificially surveyed, exoticized, and mocked.

Within the Eastern European context, it is my assertion that *The Whopper Virgins* focused particularly on Romania because of its relative poverty. For example, Borat purports to be from Kazakhstan. However, Borat's supposed 'village' was actually filmed in the Romanian village of Glod (instead of Kazakhstan) because the filmmakers claimed that they could not find a village

---

4   Bronwen Low and David Smith make almost this identical argument, claiming "Part of the problem of reading Borat as straight satire is that it is hard to separate those moments where we are (critically) laughing with Cohen, as Borat, at bigoted North Americans and North American culture and our own implication in this bigotry and culture, and where we are laughing at Borat as himself." (2007, 31).

5   For example, the United States displayed the people of the Philippines in 'human zoos' following the war in the Philippines. See Stănescu (2012, 69–100).

in Kazakhstan poor enough to meet with the stereotypical images they desired for the beginning of the movie. In fact, like the case of *The Whopper Virgins*, the Romanians in *Borat* were told that they were being filmed for a documentary when, in reality, they were being filmed purely for comic value (BBC 2008). The point is that these minstrel white performances do not simply show white stereotypes; they always show *poor* and rural white performances (that is, they are not depicted as sophisticated urbanites).

As Anikó Imre and Alice Bardan have argued (reflecting on *Borat, The Whopper Virgins*, and a Folgers ad which showed Romanians unable to figure out how to make instant coffee):

> These examples identify Romania, more than any other postsocialist country, as a site of dark television and movie tourism, the last vestige of grit and shock on which Western media producers and audiences can draw to reinvent the horrors of the Cold War as commercial docu-kitsch, which, at the same time, is the authentic condition of a strange, primitive people descended from a vampire count.... As in the aftermath of *Borat*, Romanians' outraged, and, in some cases, traumatized reactions to such representations rarely left Romania. (2010)

As DuPuis' earlier claim makes clear, the stereotype of the 'effeminate rice eater' has always operated along race as well as class lines. DuPuis' argument is that white class anxiety was articulated (and then waylaid) via rhetoric of meat, class, and masculinity. I argue that the very effectiveness of *The Whopper Virgins* relies on the same logic. The sort of implicit argument seems to be that *even white men* (like the white working class men before them) run the risk of becoming 'effeminate [Asian] rice eaters' if they do not learn to consume enough Western style meat. Carol Adams has discussed, in some detail, how modern day fast food advertisements for meat rely on ideas about masculinity among their viewers. For example, in a different ad, Burger King appropriated the lyrics from the traditional feminist anthem, "I Am Woman"; replaced them with "I Am Man," and linked the theme with eating enough Whoppers (Burger King Corporation 2007).[6] Carrie Freeman and Debra Merskin's have documented this same bias in their own survey of gender norms within fast food advertising:

---

6 See also Candy Sagon, "He Eats, She Eats," *The Washington Post* (Washington, DC), June 7, 2006, http://www.washingtonpost.com/wp-dyn/content/article/2006/06/06/AR2006060600304.html.

> Ecofeminism has critiqued this patriarchal domination of animals and nature as being linked to sexist oppression of women, contributing to a larger environmental and animal rights discourse that seeks to reduce humanity's role in the destruction of other species. Yet, despite these critiques, the masculine identity of man as defined by meat-eating is still celebrated by media in the twenty-first century, particularly in fast food advertisements. (278)

I believe that a similar logic is deployed in *The Whopper Virgins* concerning white American anxiety about race, nationality, class, and their linkages to "manhood."

These insights are underscored by the linkages to 'Transylvania' and, hence, 'Dracula.' Katrien Bollen and Raphael Ingelbien argue that the image of the 'vampire' has also always been about the fear of exotic otherness, immigrants from Eastern Europe, racial 'passing' and, most of all, fear of the resulting 'blood mixing' (2009). Dracula physically emigrates to Britain; he 'mixes his blood' with the population in a way that poisons the population, and Dracula's plan for achieving his conquest of Britain is based on 'polluting' its women. There is also the possibility of a certain level of anti-Semitism in the reference to 'Transylvanian famers.' As Sara Libby Robinson has argued, while *Dracula* was based on an overall fear of white racial 'passing,' the legend particularly focused on a fear of the Jewish immigration from Eastern Europe (2011).[7]

The Whopper Virgins popularity represents a domestication (and mockery) of the 'Asian', 'Eastern European' and possibly 'Jewish threat.' On the one hand, the (presumably Anglo-Saxon Protestant) audience is reassured, as the documentary shows the so called 'model minorities' as being so unintelligent that they cannot figure out how to eat a hamburger; on the other hand, this supposed deficiency is remedied by their 'conversion' to an American style diet of Western meat and fast food. Like the idea of the effeminate rice eater before it,

---

7 Allan Nadler, building specifically on Robinson's work, put the same argument:

"Dracula's features are 'stereotypically Jewish ... [his] nose is hooked, he has bushy eyebrows, pointed ears, and sharp, ugly fingers.' As for his behavior, Robinson situates Dracula in the realm of fin-de-siècle national chauvinism, which viewed non-Anglo-Saxons—and Jews in particular—as dangerous interlopers loyal only to their alien tribe. 'Like many immigrants, Dracula has made great efforts to acculturate himself to his new country and to blend in with the rest of the population, through studying its language and customs ... [his] greatest concern is whether his mastery of English and his pronunciation would brand him as a foreigner.' Likewise, Stoker mines anxieties over Jewish dual loyalty. 'The one identified person whose aid Dracula enlists in escaping Britain is a German Jew named Hildesheim, 'a Hebrew of rather the Adelphi Theatre type, with a nose like a sheep,' who must be bribed in order to aid Stoker's heroes." (2011)

what *The Whopper Virgins* advertisement actually does—and what I believe to be the source of its extreme popularity—is to waylay fears about immigration and cultural interaction via the argument of dietary change. If the people of the world, the documentary seems to argue, will simply eat like 'Westerners,' then they will 'become' Westerners.

The Guy Who Gave Kevin a Coat

The widespread appeal of *Whopper Virgins* to an American audience lies in the continuation of an argument, formerly made during colonization, that non-Western countries lack appropriate levels of 'masculinity,' virile 'willpower,' and technological innovation, because they do not eat enough Western style meat. The Burger King ad series repeats the tropes of the 'effeminate rice eater' in its implicit commentary on the evolving relationship between humans and animals in terms of globalization and technology. In the first place, the documentary (exclusively produced and aired online) only becomes possible because of emerging technologies in which television advertisements do not so much serve the purpose of promoting a product as of introducing much longer ads that can then be watched online. Taken alongside the ascendancy of 'reality television' and the increasing popularity of documentaries, Burger King's ad campaign seems to represent the rise of 'documentary/reality television' commercials, in which the tension between information and profit which has always existed in any commercial has now (via the inexpensive production and distribution costs of releasing a 'film' via the internet) taken on a more complete blending. How many Americans now believe that there is a country called 'Transylvania' where people wear strange traditional clothes and do not know how to eat a burger? More importantly, *The Whopper Virgins* offers a message that interweaves 'technology' and 'eating hamburgers' into a single and supposedly unique Western phenomenon, contrasting this with an archaic, romanticized, and grossly exaggerated image of a technologically deficient 'Other.'

While all of these cultures discussed are omnivorous, the documentary as a whole still invokes the scientifically unsubstantiated claim that it was meat eating which evolutionarily 'created' humans by causing both humans' larger brain capacities and the advent of tool usage (McBroom 1999).[8] By juxtaposing those people who are supposedly 'completely off the grid' with

---

8 See also Pollan, Michael. *The Omnivore's Dilemma: A Natural History of Four Meals*. New York: Penguin Press, 2006, and Paul Roberts, *The End of Food*. Boston: Houghton Mifflin Company. 2008.

the viewer who is now watching this film, *The Whopper Virgins* documentary conveys the not-too-subtle suggestion that there exists some unspecified linkage between 'technology' as a whole and eating of the 'uniquely American' burger. This contrast is most clearly expressed in the images of people (largely produced, it appears, via selective film editing) not even being able to figure out how to eat a hamburger. In this move, the hamburger itself becomes mechanized as almost a type of machinery and technology which the hopeless 'Whopper Virgins' cannot operate until the Western filmmakers graciously explain its 'operation.' This argument is remarkably counter-intuitive because to believe that someone would not be able to figure out how to eat a hamburger, we also have to suppose that that this person had never even seen a sandwich (or any of its culturally related correlates) before. Yet, while counter-intuitive, what this ad does is elevate the hamburger to become precisely the opposite of what in reality it represents; instead of a symbol of cultural homogenization and uniformity, it becomes uniquely American, the essence of 'our' shared culture. At the same time, instead of representing the homogenous practice of imposing American dietary habits throughout the entire world, the hamburger becomes a site of inter-cultural sharing and an object which is (we are told by the filmmakers) uniquely desired by the isolated communities who "want to meet new people and try new things." And instead of a ubiquitous and uninteresting remnant of technologically produced uniformity, the burger (via the lens of the documentary) becomes itself exoticized as a fascinating and complicated foreign food which people throughout the world have never seen before and could not even comprehend.

The documentary *The Whopper Virgins* additionally exploits the cultural stereotype linking technology and masculinity through its very name—The Whopper 'Virgins'—which, in a single term, masterfully interweaves sexual inexperience, meat-eating (which, as Carol Adams has shown, is itself already highly sexualized),[9] and the connotations of incompleteness and supposed desire for the new, a theme of virginity which the documentary underscores with its continual repetition of 'purity,' i.e. "the purest taste test in the world." Hence, the peoples of Romania, Thailand, and Greenland are simplified and romanticized into a state of child-like innocence, purity, and virginity, as the supposed inability to eat a hamburger renders adults of other countries into clumsy children, for American viewers. The documentary filmmakers may, at times, even claim to appreciate this sense of purity and Eden-like bliss. For example, in one scene, an unnamed man in Romania gives the filmmaker a coat for no compensation that we are told had taken a month to make

---

9   See Carol J. Adams *The Sexual Politics of Meat: A Feminist-Vegetarian Critical Theory*, 1990.

(*The Whopper Virgins* 2008). And yet, at the same time, in these dual images of 'virginity' and childhood there remains the suggestion that these peoples and their cultures are incomplete. Hence, like a child, the film invites us to appreciate the people of other cultures and dietary practices as sweet, cute, and gratuitously generous in a childlike manner and yet, in a very real sense, still lacking in personhood. For example, although we are told the names of multiple members of the documentary crew, *not a single person from any of the non-Western countries is ever named*. This produces such strange moments as the documentary credits thanking "The guy in Budesti [sic] who let us run an extension cord from his house and loaned us a hammer to fix the propane tank", and even "The guy who gave Kevin a coat." The documentary suggests they are incomplete people: incomplete in terms of technology, incomplete because of their social isolation and, ultimately, incomplete in their dietary practices.

An Un-Happy Meal: Fast Food as Humanitarian Assistance

The filmmakers clearly suggest this idea of incompleteness, as well as the solution of eating Western style meat, when, after the taste test, they take burgers into the participants' villages themselves. I find this part of the film the most disturbing because it clearly belies the claim that the actual motivation was a taste test, or even a publicity stunt tied to a taste test, since no taste testing occurred in these villages at all. Instead these actions cinematographically take on, I believe intentionally, all of the trappings of a humanitarian relief mission.

For example, we are shown the man who grilled the burgers, after being flown in by helicopter, using a large forest-covered netting to give away food. The scene is drawn from the set of images in which food drop-offs are almost universally displayed—an iconography of humanitarianism which is underscored by the repeated scenes of suggested 'poverty,' contrasted with the repeated actions of the filmmakers. The filmmakers further blur the line between humanitarian relief and commercial exploitation in that Burger King undertook actual humanitarian actions right alongside the 'sharing' of the burgers, such as giving away free toys to children. After all, is not the coupling of hamburgers and 'free' toys for children the essence of the appeal of the "Happy Meal"?[10] Hence, Burger King's press release after the event can interweave the

---

10   A "Happy Meal" is the name given to the children's meal at McDonalds. It usually contains a 'free' toy for the child. Burger King offers a similar correlative under a slightly different name.

FIGURE 5.3 *Burger King's humanitarian mission to deliver Western meat to non-Westerners.*

giving of the burgers and the giving of toys into a single "giving back to the community that Burger King always shows" (Burger King Corporation 2009).

There is also a final dimension to each of these displays which Burger King uses to shelter this production from any criticism—that of humor. The implied suggestion is that what we are watching is supposed to be funny. Hence, the documentary is both a 'documentary' and at the same time, via an unspoken implication, understood not to be a 'real' documentary. As media savvy viewers, we are simultaneously supposed to believe in the research that we are seeing and at the very same time understand that what we are witnessing is a little contrived and funny, a little tongue in cheek, and it is this humor, along with the humanitarianism, which is supposed to ward off all criticism. To criticize this documentary, particularly in a methodical and precise academic manner, is therefore to suggest that somehow we failed to 'get the joke.'

This pseudo-humanitarian assistance, coupled with the implied humor, is what I find most distressing, as many, if not all, of the communities shown (and certainly Maramureş) suffer from real poverty, including hunger, and could in fact benefit from actual, sustained food relief. Instead their poverty is displayed as only a form of amusement and profit for the Burger King franchise. We are invited, quite literally, to laugh at these diverse peoples of the world, their different dietary practices, clothes and cultures, and their inability to understand 'complicated' American manners and food. Much like the ministerial shows before them, the ads present a carnival or fair-like atmosphere of both amusement and fake information in which people from around the world are displayed solely for profit. This humor was underscored by a *Saturday Night Live* skit parodying these same advertisements. In one scene, a woman explains via a supposed 'translator': "She says she is not a virgin... It was her uncle," or, in another scene, "He says he will agree to say anything you want if you will let him bring this food home to his village" (*Saturday Night Live*, 2009). I understand that the purpose of the *Saturday Night Live* skit is to poke fun at Burger King and not towards Romains. Yet, I have never discussed this *Saturday Night Live* skit in class without most of the students bursting out in laughter. And, every time I hear this laughter, I feel a sense of discomfort. Much as with the example of *Borat*, it is impossible for me to parse to what degree my students are laughing at the critique and parody and to what degree they are still simply laughing at the continuing comedic images of the 'childlike' Romanians.

It is in this context that I would like to reconsider the final scene of *The Whopper Virgins* ad: the scene of the helicopter arriving as a humanitarian relief mission bringing hamburgers. I believe, and the comments of the fans of the ad support, that the giving of not just food, but American meat, was itself the 'humanitarian' aspect of the mission that was supposed to help transform the childlike Romanians into full humans capable of enjoying 'human' rights. What concerns me here is the suggestion that what the Romanians (as well as all other starving peoples of the world) most need is meat, and specifically hamburgers. This claim is of such particular concern because the world's overconsumption of meat is the single largest contributor to global world hunger (Motavalli, 2001). As such, *The Whopper Virgins* campaign recasts one of the largest causes of world hunger (excessive consumption of meat) as one of the 'solutions' to the problem.

Conclusion

It is my claim that we can observe certain continuities in terms of gender, xenophobia and diet between the stereotype of the 'effeminate rice eaters' of the nineteenth century and *The Whopper Virgins* advertising campaign of the twenty-first. In both cases, non-Western peoples are portrayed by the West as lacking adequate levels of Western style meat, which, in turn, is used to suggest that they lack appropriate levels of masculinity and personhood.

These stereotypes are used in turn to naturalize the poverty of the third world, not as derived from, say, market forces instituted by Western countries, but instead as derived only from the diet, and hence the culture, of the people themselves. Each of these aspects of Burger King's project works intentionally and synergistically together. It is because the documentary portrays the people of Maramureş as suffering from a lack of exposure to Western dietary practices, specifically the hamburger, as the reason for their lack of technology and relative 'poverty' that taking the hamburgers into the villages themselves becomes, within the logic of the documentary, the best solution and ultimate act of humanitarian assistance. The very actions of which fast food franchises are most guilty—worldwide standardization and destruction of all forms of dietary and cultural diversity as well as causing mass animal suffering and environmental degradation—become normalized as humanitarian assistance. Thus, the proposed 'solution,' both in the historic and contemporary example, is not one of redistributive justice or a need to globally restrict the current exponentially increasing levels of meat consumption, but to provide pseudo-humanitarian aid in the paternalist form of complete cultural assimilation to the Western norm of excessive and cheap meat consumption. After all, as the letters to Sharon Akabas made clear, Burger King is simply defending everyone's "God-given right to a juicy hamburger."

References

Adams, Carol J. *The Sexual Politics of Meat: A Feminist-Vegetarian Critical Theory*. New York: Continuum. 1990.

Baron Cohen, Sacha et al. *Borat*. Beverly Hills, California: Twentieth Century Fox Home Entertainment, 2007.

BBC. "Village 'Humiliated' by *Borat* Satire." October 26, 2008. http://news.bbc.co.uk/2/hi/europe/7686885.stm.

Bean, Annemarie, James Vernon Hatch, and Brooks McNamara. *Inside the Minstrel Mask: Readings in Nineteenth-Century Blackface Minstrelsy*. Hanover, NH: Wesleyan University Press, 1996.

Bollen, Katrien and Raphael Ingelbien. "An Intertext that Counts? Dracula, The Woman in White, and Victorian Imaginations of the Foreign Other." *English Studies* 90.4 (2009): 403–420. DOI:10.1080/00138380902990226.

Burger King Corporation. "The Whopper Virgins." Press release, 2009.

Burger King Corporation. *Burger King: A Deschis Restaurantul Cu Numărul 8*. Press release. July 1, 2012. http://www.burgerking.com.ro/.

Burger King Corporation. "I Am Man." January 26 2007. Accessed May 28, 2014. http://www.youtube.com/watch?v=vGLHlvb8skQ.

Business Review. "McDonald's Romania Celebrates 15th Anniversary." June 21, 2010. http://business-review.ro/news/mcdonald-s-romania-celebrates-15th-anniversary/9432/.

Caiozzo, Paul. Personal website. Accessed June 1, 2012. http://paulcaiozzo.com/awards.html.

Crispin, Porter & Bogusky. "The Whopper Virgins." *Burger King Corporation.* 2008. Accessed on June 11, 2015. www.whoppervirgins.

Corning, J. Leonard. *Brain Exhaustion, With Some Preliminary Considerations on Cerebral Dynamics.* New York: D. Appleton and Company, 1884.

Doing Business in Romania. "McDonald's Romania Invests EUR 9m in Q1 to Spread Tentacles in the Field." May 19, 2009. Accessed June 1, 2012. http://www.doingbusiness.ro/en/business-news/11742/mcdonalds-romania-invests-eur-9m-in-q1-to-spread-tentacles-in-the-field.

DuPuis, Melanie. "Angels and Vegetables: A Brief History of Food Advice in America." *Gastronomica: The Journal of Food and Culture* 7.3 (2007): 34–44.

Freeman, Carrie and Debra Merskin. "Having it His Way: The Construction of Masculinity in Fast Food TV Advertising." In *Food for Thought: Essays on Eating and Culture*, edited by Lawrence R. Rubin Jefferson, 277–293. NC: McFarland, 2008.

Imre, Anikó, and Alice Bardan. "Dracu-Fictions and Brand Romania." *Flow*, 11 (February 19, 2010). http://flowtv.org/2010/02/dracu-fictions-and-brand-romaniaaniko-imre-and-alice-bardan-university-of-southern-california/#discussion.

Laudan, Rachel. "Power Cuisines, Dietary Determinism and Nutritional Crisis: The Origins of the Globalization of the Western Diet." *Interactions: Regional Studies, Global Processes, and Historical Analysis.* February 3, 2001. Accessed Jan 16, 2010). http://www.historycooperative.org/proceedings/interactions/laudan.html.

Lindfors, Bernth. *Africans on Stage: Studies in Ethnological Show Business.* Bloomington: Indiana University Press, 1999.

Low, Bronwen, and David Smith. "*Borat* and the Problem of Parody." *Taboo: The Journal of Culture and Education* 11.1 (Spring–Summer 2007): 31.

McBroom, Patricia. "Meat-Eating Was Essential for Human Evolution Says UC Berkeley Anthropologist Specializing in Diet." University of California, Berkeley, Press Release. 1999.

Motavalli, Jim. "The Case Against Meat: Evidence Shows that Our Meat-Based Diet is Bad for the Environment, Aggravates Global Hunger, Brutalizes Animals and Compromises Our Health." *E: The Environmental Magazine*, December 31, 2001. http://www.emagazine.com/archive/142.

Nadler, Allan. "Imaginary Vampires, Imagined Jews." *Jewish Ideas Daily*. July 11, 2011. http://www.jewishideasdaily.com/921/features/imaginary-vampires-imagined-jews/.

Nederveen, Pieterse J. *White on Black: Images of Africa and Blacks in Western Popular Culture*. New Haven: Yale University Press, 1998.

Pollan, Michael. *The Omnivore's Dilemma: A Natural History of Four Meals*. New York: Penguin Press, 2006.

Roberts, Paul. *The End of Food*. Boston: Houghton Mifflin Company. 2008.

Robinson, Sara Libby. *Blood Will Tell: Vampires as Political Metaphors before World War I*. Boston: Academic Studies Press, 2011.

Sagon, Candy. "He Eats, She Eats." *Washington Post*, June 7, 2006. http://www.washingtonpost.com/wp-dyn/content/article/2006/06/06/AR2006060600304.html.

Saturday Night Live. "The Whopper Virgins," Series 34: episode 12. Original air-date January 10, 2009. http://www.hulu.com/watch/53099/saturday-night-live-whopper-virgins.

Stoker, Bram. *Dracula*. Melbourne, Vic: Brolga. 2012.

Stănescu, Vasile. "'Man's' Best Friend: Why Human Rights Need Animal Rights from the Philippines to Abu Ghraib." *Journal for Critical Animal Studies* 10, iss. 2 (Special Issue: Prison and Animals) (2012): 69–100.

Trosclair, Carroll. "Burger King's Controversial Whopper Virgin Ads: BK Campaign Raised Cultural and Nutritional Issues for the Industry." Suite101.com. Jan 1, 2009. Accessed Jan 16, 2010. http://tvadvertising.suite101.com/article.cfm/burger_kings_controversial_whopper_virgin_ads#ixzz0j3Rm8bp5.

York, Emily Bryson. "Controversy Is Just What BK's 'Whopper Virgins' Is After." *Advertising Age*, December 8, 2008. Accessed July 19, 2012. http://adage.com/article/news/controversy-burger-king-s-whopper-virgins/133063/.

York, Emily Bryson. "'Whopper Virgins' Rivals Online Success of 'Freakout'." *Advertising Age*, January 12, 2009. Accessed July 19, 2012. http://adage.com/article/digital/whopper-virgins-rivals-online-success-bk-s-freakout/133721.

CHAPTER 6

With Care for Cows and a Love for Milk: Affect and Performance in Swedish Dairy Industry Marketing Strategies

Tobias Linné and Helena Pedersen

Prelude: An Afternoon in the Countryside

When we arrive, half an hour before the start of the event, there are already people walking from the field temporarily turned into a parking lot up the small hill slope to the farm. We park the car and begin making our way up to the farm. As we approach it, the narrow road is getting more and more crowded with people. There are school children in yellow vests with their teachers, elderly people in wheelchairs with their assistants and families, mothers, fathers and children of different ages. Groups of teenagers, some from the agricultural school that the farm is affiliated with, are laughing, shouting and screaming. Along our way large signs and banners hang, advertising *Skånemejerier*, the dairy corporation behind the event. By the signs tables have been arranged where visitors are offered free chocolate flavoured milk along with a cinnamon bun. The queues in front of the tables are long with children eager to get some free food and drink.

Suddenly we have to move to the side as we hear a truck approaching from behind us. It is driving slowly and carefully up the road. On the back of the truck, three people are standing, tampering with a speaker system. Soon country music, a big Swedish hit song from the mid-1980s, begins to flow from the speakers. It feels like a carnival: the music, the sun shining, hundreds of people talking and laughing, spreading out blankets and eating picnic food.

As we approach the barn where the animals are kept, the sounds of human voices and laughs are mixed with the cries of calves. When we get nearer we see a small enclosure, 2×2 meters, in which two newborn calves have been placed. The calves cram together in the middle of the enclosure, looking around at what is going on. They seem stressed about the attention they are getting from children standing along the fences trying to pet them. Right next to the calves is another enclosure, where a family of sheep is showcased. The sheep seem somewhat more comfortable with the situation; at least the lambs are there with their parents. The calves are by themselves, disoriented and frightened.

There are no signs telling us anything about the animals, no representatives from the farm there, just the animals behind bars in the middle of a lawn with hundreds of people running around them, looking, pointing, laughing.

We stand there for 20 minutes, listening to people talk about their weekend plans and watching the place getting more and more crowded. The best places along the fence are filling up fast. Suddenly the music is silenced and a voice is heard in the loudspeakers; it is the headmaster of the school that runs the farm. "Welcome everyone to our school on this beautiful day, and to our annual release of the cows". In her speech, the headmaster especially addresses the children as she talks about how important it is to learn where food comes from. She mentions how many litres of milk a cow produces every day, and that the cows live for approximately seven years (she leaves out the fact that they could have lived to at least 20 years of age had they not been exploited for their milk and killed).[1] She continues by saying that after the cows have been released into the pen we are welcome to take a look inside the barn to see what it is like for the cows in there. "Now I can hear that the girls are really excited to come out" she says, pointing towards the barn from where an indistinctive sound is heard. "They have been standing inside all winter, and are really really longing for some running. So help me with the countdown now; ten, nine, eight, seven, six, five, four, three, two, one, open the doors!"

The doors open, the audience gasps, and then the cows start running out. Some are running fast, straight into the enclosure and as far away as they can get. Some run for a little while, then stop, turn around and look at the other cows, and start running again. Others walk rather than run, as if their legs are too weak after having stood inside a barn for more than half a year with little or no exercise. A woman beside us is talking to her friend about how the cows must feel running out in front of so many people. "Maybe they don't want to come out" she says, "maybe they are afraid and feel uncomfortable in front of so many people, I know I would be". Some of the cows seem rather uncomfortable with the attention from the more than 2000 people that have come to look at them. Some even refuse to leave the barn, and have to get pushed out by the staff.

Within a few minutes the cows are all out of the barn and walking around in the pen. People start to make their way back towards the cars. As we leave we

---

1   See the next chapter in this volume, Melissa Boyde's "Peace and Quiet and Open Air": *The Old Cow Project*.

get a last glimpse of some of the cows that are approaching the fences where people are taking pictures of them.[2]

Aims and Objectives

In this chapter we explore the marketing strategies of the Swedish dairy industry, especially the so-called pasture releases and open farm events at Swedish dairy farms that have been a great success during recent years. These events are presented to the guests as learning opportunities about milk production, farm life and the animals' everyday lives. Families with children and school classes are targeted for the events, and the latter especially invited. The activities are designed to let the children follow the way of the milk, from the moment of arriving at the farm where calves newly separated from their mothers are showcased, to the end where visitors are invited to taste the final products.

Our chapter will give a guided tour through these events, exploring how the pasture releases and open farm events embody, shape and legitimize certain values and ideals of human-bovine relations. The analyses follow two trajectories of scholarship in critical animal theory. The first of these trajectories draws on Foucauldian analyses of the production of farmed animal subjectivities in meat and dairy industry settings (Cole 2011; Holloway 2007; Thierman 2010). The second is related to what has been called the 'new carnivore' movement (Gutjahr 2013; Parry 2010; Potts, Armstrong and Brown 2013) and the 'happy meat' discourse (Cole 2011; Gillespie 2011; Stănescu 2014), referring to the frequently expressed idea that meat produced at small scale organic farms, where the animals are slaughtered more 'humanely', represents a more ethical way to consume animals, than the consumption of industrially produced meat.

We argue that, in the pasture releases and open farm events, the production of bovine subjectivity and new carnivorism/happy meat should not be seen as separate phenomenon; rather, they are intimately interrelated by educational elements relying on bovine emotional labour. These educational elements take on certain performative dimensions (zooësis) that are integral parts of the success of the new carnivore and happy meat regimes.

---

[2] The opening to this chapter describes a public so-called pasture release event, taking place on the 23rd of April 2014 at an agricultural school in Svalöv, a rural municipality in the southernmost part of Sweden. The event was attended by Tobias Linné together with two animal rights activists and film-makers working on a documentary project on the living conditions of cows in the Swedish dairy industry.

The chapter relies on two interconnected methodological approaches. First, ethnographic field studies and observations of pasture releases and open farm events were carried out. Between spring 2012 and spring 2014, six different farms were visited on nine separate occasions, during both pasture release events and open farm events. The events were visually documented by camera and the images were analyzed using an open-ended qualitative approach focusing on reoccurring themes relating to the values and ideals of human-bovine relations represented. Twenty-one brief on-site interviews were also conducted with visitors to these events. Second, semi-structured interviews were conducted with six farmers on the respective farms, as well as two interviews with representatives of the Swedish dairy industry and two interviews with representatives of the Swedish dairy lobby organizations *Swedish Dairy Association* and the *Association of Swedish Dairy Farmers*. In addition, printed promotional material from the Swedish dairy industry has also been gathered at the events and analyzed.

Before engaging in more in-depth analysis of the pasture releases and open farm events as sites that shape and legitimize certain human-bovine relations, we first provide a brief overview of the Swedish dairy industry and its marketing strategies.

The Swedish Dairy Market and Industry Marketing

The Swedish dairy industry is the most economically significant branch in the Swedish agricultural sector, accounting for a fifth of the total production worth in 2013 (Statistics Sweden 2013). During recent years, the Swedish dairy industry has become increasingly intertwined with the global animal economy. Merging, where larger dairy corporations have taken over smaller ones, is a strong trend (Jönsson 2005), as is the replacing of smaller farms with larger ones. The number of farms delivering milk to dairy processing plants has for example decreased by half since 2002 up until 2013 (from approximately 10 000 farms to approximately 5 000 farms), while the average number of cows per farm has increased from 42 to 71 cows per farm. There has also been merging with global dairy corporations and increases of the export of milk and dairy products in recent years. For instance, the export of non packaged milk increased from 37 000 tonnes in 2009 to 67 000 tonnes in 2013 and the export of non packaged cream from 924 tonnes in 2009 to 5774 tonnes in 2013 (Statistics Sweden 2013).

The two largest dairy corporations, *Arla* and *Skånemejerier*, have both entered the international milk market, beginning to sell their milk to new developing markets (such as China and India) and merging with other dairy companies from multiple other countries (Swedish Board of Agriculture 2012; Ekoweb Sverige 2013). During the same period, the Swedish dairy industry has also gone through a crisis. The crisis, a result of declining milk consumption in Sweden and an increasing international and national competition of market shares, has led to a decrease in the number of dairy farmers by 50 percent during the last ten years (Statistics Sweden 2013). Fears have been raised that in a few years Swedish milk production might be outcompeted altogether (Swedish Board of Agriculture 2012; Jönsson 2005). These tendencies all provide a background to the importance for the dairy industry to promote itself to Swedish consumers through pasture releases and open farm events. There is a need for a nostalgic counter image to the processes of globalization, rationalization and effectivization that the industry is going through. The release of the cows onto pasture provides an image of a local and organic dairy production. It responds to larger societal debates about sustainability in the production of food and to the demands of concerned consumers (Stănescu 2014).

According to Swedish animal welfare legislation, dairy farmers must let their cows out on pasture for at least six hours a day for two to four months (depending on where in Sweden the farm is situated) during the period of April 1st to October 15th (Swedish Board of Agriculture 2014). For the remaining part of the year, the cows are kept inside barns in different housing systems.[3] The legislation has often been pointed to when the dairy industry is trying to promote itself as environmentally and animal friendly. During the last decade, the pasture releases have been increasingly framed as entertaining events, directed to the public and presented as a perfect summer outing for the whole family. Starting off with 50 or 100 people in the audience the pasture releases have grown and are now given much attention in the dairy industry marketing efforts (on webpages, social media and in advertising dispatched to schools). During the spring of 2013 an estimated 140 000 people attended the pasture releases at the Arla dairy farms alone, and the events that we visited all attracted more than 2000 persons each (Arla news 2014). Apart from using the events in their image and brand building on webpages and in social media, dairy company representatives are also present at the pasture releases, offering visitors free milk among other things.

The open farm events bear many similarities with the pasture releases. At a number of dairy farms, many of which also organize pasture release events, it is possible to "come and say hello to the cows" and watch them getting milked

---

3   Approximately 55 % of the cows are kept in free stalls, and 45 % are held tied up in the stalls (Dairy Sweden 2014).

(Skånemejerier 2010). The open farm events are especially offered to schools that have the opportunity to book a farm for a whole day to see how milk production is carried out (Arla minior 2011a). Usually however, these events are open to the public, and like the pasture releases they are attracting many thousands of people out to the country to have a look at the cows and the farmers' work.

The pasture releases and open farm events continue a tradition of dairy promotion that in Sweden is almost 100 years old. The dairy industry has historically occupied a special position in Swedish society, and is still often not viewed as other commercial actors, but rather like a semi-authority in which people report to have very high trust (Jönsson 2005; Sydsvenskan 2012). One example of this is the fact that dairy companies are allowed to display advertisements in school facilities where other commercial actors are banned. In addition, schools and pre-schools receive teaching and learning materials from the dairy industry, materials that have been tailored to fit with the school curriculum. Teachers are presented with complete pedagogical plans to fulfil learning objectives. A visit to a dairy farm is often an element in these plans (Arla minior 2011b).

"The Way of the Milk": Short-Circuiting the Dairy Production Process

We start "The way of the milk tour" at the open farm outside one of the barns. In a small temporary confinement two young calves have been placed. They have been separated from their mothers just two days ago the farm employee informs us, as she is bottle-feeding one of the calves. Children are invited to go into the confinement and pet the calves whilst the woman is talking about the handling of the milk, how often the trucks come to the farm to pick it up, and how much milk is produced every day. She is using a neutral and technical language; it is presented as a process of logistics and automation, as if it were machines producing the milk. No violence, no conflict, despite the fact that the whole process is a result of a forced separation of calves from their mothers.[4]

---

4 To produce milk, cows need to give birth to calves. Therefore, the cows of the dairy industry are inseminated about once a year. Normally, the calf is taken from the cow either directly or 2–3 days after birth. Cows are maternal animals and both the mother cow and the baby calf suffer from being separated at such a young age (Lidfors et al. 2004). Half of the calves born

Later on during the tour we are shown how the milking machine works. What is striking about this presentation of the milk automation system is how little attention is actually given to the cows as animals, despite the fact that they are central to the production of milk. It is the milking machine and how it works that is the prime focus. The animals seem secondary to the process, and we only get a small glimpse of the cow being milked from our view in the control room where we watch a dairy industry representative present the milking system.

The relationship between the cows and the milk taken from them is reimagined at the open farm events. The visitors are offered the possibility to watch the cows being milked, but this view is controlled in various ways, and the visitors guided in their interpretation of what is going on. During the open farm and pasture release events the cows are mechanomorphically referred to as 'milk machines' when presented to the children (cf. Crist 2000). Similarly, in the interviews, one of the farmers uses the image of the cow as a factory and a machine for yielding an economic gain:

> They are like a living biologic factory, it is like you put something in one end, grass, and out comes this fantastic product, that is both cheap and nutritious, one of the best things you can drink [...] And I think that our cows, when you have them on pasture like this, the animal lasts longer too, and then you get a much better economic result than if you have to send it to slaughter prematurely. [Farmer 4]

At the entrance to the barn where visitors are invited to enter and look at the cows being milked a milk automat has been placed. The idea seems to be that visitors can have a glass of milk before they enter the barn to look at the cows being milked. The milk is of course not directly from the cows at the farm, but has been produced at some undisclosed dairy farm, then sent to the factory processing plant and subsequently sent to the farm. By the milk automat in front of a milk barrel and alongside some boxes of milk, bundles of green grass and clover have been placed.

This presentation of processed milk in automats and the way the milk is symbolically connected to the green grass, represents a mix up of the production cycle and hence an educationally questionable version of the way of the

---

are male and since they do not produce any milk most of them are killed for meat at about 18 months of age. The female calves are either selected as dairy cow replacements for their mother, or sent to slaughter (Svenskt Kött 2014; Arla 2012).

milk. The reality of the process by which cows are farmed and bovine lactation is managed is short-circuited. Nowhere on the tour are we informed about or allowed to see the separation of the calves from their mothers, the forced impregnation[5] of the cows or animals being sent to slaughter.[6] The open farm presents the milking process to the visitors reimagined as idyllic and happy; much like how the lives of animals in zoos are presented to give the visitors a positive experience. The way of the milk is reduced to a simplified, clean and morally unproblematic process of grass in-milk out, hiding the exploitative practices at work when the cows 'give' their milk to humans (Molloy 2011).

At the final stage of the way of the milk tour, when we have arrived at the yard in front of the farmer's mansion, dairy industry representatives have put up tents and offer free dairy products for all visitors. In one of the tents, two cooks are performing a cooking demonstration with dairy products. Walking around the yard is a human dressed up in a black and white calf costume with a giant smiling head on top. This false calf with a human inside is Kalvin, the Skånemejerier company mascot.

Again, the production cycle is short-circuited; the milk seems to have just passed from the barns and the milking machines into these final products, with little mention of the processing plant that the milk has to go through to be made into the 'delicacies' that the visitors taste in the tents. Cutting out the plant from the production cycle underscores the naturalness of the milk, while paradoxically at the same time the animals themselves are represented as machines and factories.

Animal to Drinkable:[7] From Bodily Fluids to Chocolate Drinks

To follow the tour of the milk at the open farm is to witness the transformation of animal to commodity. At the beginning of the tour the milk-baby animal connection is highlighted in a morally neutral way and the animals are the centre of attention. Then gradually, the animals and the animal origin

---

5 The majority of dairy cows in Sweden are impregnated by artificial insemination, from when they are about 15 months of age. Bulls are used for breeding and a single animal can father many thousands of calves each year by artificial insemination (Djurens rätt 2013).

6 The dairy industry and the meat industry are closely connected. Approximately 65 % of the total amount of beef produced in Sweden comes from either 'spent' dairy cows who are not producing enough milk to be economically efficient, or from their offspring (Svenskt Kött 2014).

7 With this subtitle we paraphrase Noëlie Vialles (1994).

of milk are erased. The milk is turned into a cultural commodity. The cooking demonstrations using dairy products at the end of the open farm tour is one example of this transformation. Another poignant example is that both at the pasture releases and open farm events, children are given chocolate milk to drink. Instead of a drink that tastes and looks like the milk coming from the cow's body, they are offered a culturalized version of milk, a product that tastes quite differently, and does not even resemble milk in colour. At the end of the way of the milk tour that started with real calves, we also meet Kalvin, the company mascot, telling children it is okay to drink milk, thus redirecting the consumers' affective sentiments from real animals to a fake calf (Stewart and Cole 2009).

While in one sense the pasture releases and open farm events are hiding the real animals, transforming them into the absent referents (Adams 2010) of dairy production, these events are also connected to the new ideals of visibility and transparency with regard to the production of food that are central to new carnivorism. These discourses function as a refutation of the critique against animal farming. One of the farmers explains how she sees the open farms as an opportunity to argue for her way of working with dairy production:

> I love this chance to meet the consumers, and argue for my sake, my way of doing things [...] people come up and talk to me, and the children are here asking questions and we explain things to them, about the machines and the animals. And we want to show what our everyday situation is like, what life on the farm is like, and I think that is the most important part of it. [Farmer 3]

These discourses also ensure that consumers of animal products cannot be accused of hypocrisy for being disconnected from the reality that brought animal food products to their table. This is one of the core themes of the pasture releases and open farm events. One of the interviewed farmers explains:

> We know what the consumers want, and what they want is honesty, because many doors to food production are closed, people don't have a chance to see what it is like, and we have had this discussion too, how much can we really show? Maybe the consumer will be shocked because they don't know what the production is like for real? [Farmer 2]

While the open farms and pasture release events seem to be about making the dairy production visible, this visibility also functions as concealment. Partly it seems to have to do with the potentially disturbing effects the witnessing of what the dairy production really looks like might have on people. Instead of presenting a real image of cows' lives in the dairy industry, the pasture releases and open farm events continue what Jönsson (2005) describes as a fetishization of the cow in dairy industry marketing. The cows are being used to evoke a certain image of reality by simulating it, representing a yearning for the authentic, the pure and the real, an idealized version of an agrarian context that has never actually existed. The happy cows running onto green pastures are assigned visibility and subjectivity for the purpose of communicating their happiness to the people watching them. And the visits to the cow stalls are mostly, it seems, about refuting what is believed to be a false image of how dairy production is carried out, as one of the farmers explains:

> And so they can also see that the new stables are nothing like what they used to be, dark and without windows, and with bad air, but if you go to the new stables, it is light, open and, it is almost like being outside, almost like you wouldn't need to let the cows out, so the cows are much more happy, and they can be more together with their calves. [Farmer 2]

A Fantastic Spectacle: Bovine Emotional Labour

When talking to people during the pasture releases many mention how good they feel watching the cows, how they are becoming less stressed and learning to live more in the moment. The excerpt below taken from Skånemejeriers' webpage shows how the pasture releases are marketed to cater for these feelings, how they function as entertainment events, and how the cows are the stars of these events:

> Just like us, the cows are yearning for the sun, the scents and the green grass. And they know when it is time to be let out. So when the dairy farmer opens the cowshed door, they take off. The cows jump and dance of happiness. Some rush with speed all the way out to the green pastures. It is a fantastic spectacle. Come and visit one of our pasture releases. We guarantee that you will feel the spring in the air. (Skånemejerier 2013)

As the involuntary stars of these events, the cows are acknowledged as subjects as they transcend their species-being; more specifically, if they attain human-like qualities, and when they are anthropomorphized as "characters" (Stewart and Cole 2009). The excerpt from the field notes below provides an example of this:

Just outside the barn is a sign that reads "The cowwalk", playing with the appearance of the whole scenery as a catwalk and then some images of the cows accompanied by their names. "Look at that one, she looks just like you", shouts one of the onlookers in her teens while laughing at her friends. "No, that is you and that is me" the friend replies jokingly pointing at the images.

The cows are not the only entertainment at the pasture releases and open farms. The events that we visit offer a bewildering array of other activities, such as live music, quiz walks, petting zoos, and rides on a mechanical bull (an activity reminiscent of cruel rodeo entertainment), all transforming the scene into something resembling a carnival. At the pasture releases and open farms, the cows are made to work for the industry that exploits them not only as food animals producing milk. Being used as entertainment, the cows are rendered *affectively useful*. This is underlined by our observations where people are cheering, laughing and shouting in awe looking at the cows. We argue that the cows' affective usefulness constitutes an articulation of bovine emotional labour (cf. Hochschild 1983), and that emotional labour turns the cow commodity herself into a performer; a spectacle (cf. Debord 1983). This bovine catwalk becomes, in the pasture release and open farm events, an arena of *zooësis*. Una Chaudhuri (2007a) has proposed the notion of zooësis in order to conceptualize how human culture uses 'the animal', as trope and as body, in meaning-making practices. Zooësis does not only refer to the entire Western tradition of animal representations, but also to all culturally contingent human-animal practices such as dog shows, zoos, animal experiments, hunting, meat consumption etc.—each practice carrying its own history and its own repertoire; its own actors and its own audience (Chaudhuri 2007b). The pasture release and open farms events are, in this analysis, another manifestation of carefully choreographed zooësis, with its own history, its own repertoire, its own actors (the cows), and its own audience (ourselves and the other visitors with whom we share this experience). It is the affective quality of bovine zooësis that has such powerful educational appeal in these events.

The dairy industry takes great interest in how the spectacle of the pasture releases and open farms represents a possibility of emotional connection between humans and cows, a promise of an interspecies encounter with humans in harmony with animals and nature. One of the dairy industry interviewees claims that the perceived similarities between bovines and humans as experienced when watching the release of the cows—an experience apparently facilitating anthropomorphic identification with them—is key to the success of the pasture releases:

> I think the reason is that link back to nature, we need to have animals surrounding us, we need to have something like this to balance our stressed everyday life that gets more and more stressful all the time ... we have to start breathing and we have to start caring more about the food we eat and the animals [...] so I think we can relate to this, that the cows need to go on pasture these months, just like we need a vacation sometimes, just like we need the sun. [Dairy industry representative 2]

Although the events can be understood as a kind of light entertainment, they also carry a spiritual dimension. One of the farmers tells about the emotions that the pasture releases bring out:

> Cows are special animals, if you have ever looked into the eyes of a cow, you know, it is like, they have a special wisdom [...] And this is part of that, taking care of nature, having animals around you... for sure, it is also about spirituality, we have no religion any more so that's why I think nature is becoming more and more important as a way for people to heal, so it is contemplative, it is like yoga. [Farmer 3]

While the cows, according to the farmers, evoke emotional responses in the people watching them being released (an emotional response both recruited and exploited by the industry to promote its self-image of caring for the animals' wellbeing), this form of visual consumption does not actually bring 'us' closer to 'them'. Instead of bridging the species gap, and giving a deeper understanding of the cows, humans are further distanced from the animal (cf. Desmond 1999). The pasture releases and open farms reinscribe the very same human-animal species boundaries that they on a superficial level seem to be challenging. During the pasture release events, the animals are *almost human, but not quite* (Pedersen 2010). So, on the one hand, the events rely on humans identifying with the cows, seeing their emotions reflected in the presence of cows. On the other hand, the pasture releases position the cows as different from humans. The visual arrangement of the open farms and pasture releases is organized as an agri-industrial zoo or circus, with the occasional information sign by the fences telling us strange facts about the cows. This exoticizing of domestic animals further underscores the moral logic of a speciesist sociocultural order. One of the farmers talks about the amazement of the strange creature that people experience when coming to the pasture releases and open farm events:

Many of the kids coming here, and many of their parents too it seems [laughing] they have never seen a cow before, they don't even know that... They are totally amazed, the children coming here, "Oh, they are so big [the animals]" and they can pet them, and I ask them [the children] "How often do you think we milk the cows?" and they reply "Once a month". [laughing] [Farmer 1]

The ironic outcome of this pseudo-intimacy with cows in the countryside is that the human separation from the cows is further reinforced. This is also emphasized through the more educational parts of the pasture releases and open farm events, which are explored in the next section.

Educating the Masses: The Pasture Releases and Open Farms as Arenas of Child-Animal Relations

As we stroll around the open farm listening to people's conversations, it is evident that the events present an opportunity for parents to educate their children about animals and farming practices. This is also officially one of the ideas behind the events, according to one of the interviewed dairy industry representatives:

We want it to be an occasion where children can learn about nature, to care for nature and to feel responsibility for nature and for the animals. And together with LRF [The Federation of Swedish Farmers] we invite all the schools in the region to the farms, and then we have a schedule so that when the schools arrive we show them around according to this schedule. And we have a very good learning material about "Life in the countryside". It tells about what you eat and what's the difference between a heifer and a cow... and food has become an important topic, what we eat, that milk is nutritious, and some teachers say that they need to learn more about this, and so they come here, and when they do, they get educated here so that they can pass it on to the children. [Dairy industry representative 1]

However, despite the focus on the care for nature and the animals that the interviewee talks about there are hardly any ethical issues brought up during the events. Neither is there any problematizing of the dairy industry's environmental impact. These educational events are focused on *anything but* ethical and/or environmental problems. By the closed confinement of the calves at one of the pasture releases a sign is posted with information about the cows and dairy, which in tone and style is typical for the kind of knowledge offered at these events. It is the anthropocentric neutral knowledge of the human observer over the animal other, the kind of knowledge of the animal other that translates into power:

> Did you know that:
> A cow can drink more than a 100 litres of water every day
> A cow will eat on average 50 kilos of feed a day
> Cows don't like changes. The same routines and same food every day is what they like best
> A cow from our farms gives 30 litres of milk a day, year around
> A cow is pregnant for 9 months
> Heifers have their first calf at the time of their second birthday. After having their first calf, the cows start to produce milk
> A calf should have raw milk as soon as possible after its birth. The raw milk is the cow's first milk, and it is full of nutrition[8]

---

8 An alternate sign with information about cows and dairy might look something like this:
   Did you know that:
   - Cows don't have to be milked. Cows, like other mammals, only produce milk after they have had a calf. We humans have the habit of taking the calf from the cow immediately after birth and then take the milk for ourselves.
   - Most calves are taken away from their mother within 24 to 48 hours. The calf is fed milk replacers before early weaning at around 5–6 weeks. Calves would naturally suckle for 6 to 12 months.
   - There is a strong bond formed between the mother and her calf in the first few hours after birth; enforced separation is therefore a very traumatic experience for both.
   - Many cows are sick and injured. The cows are pregnant and lactating almost continuously throughout their lives. This is hard on the cows' bodies and causes many to become ill.
   - It is hard to separate the milk industry from the meat industry. The majority of all beef sold come from animals in the dairy industry.
   - Organic dairy is not much better than conventionally produced dairy. Organic cows are often outdoors more, and get more roughage, but apart from this there are no big differences between organic and conventional production of dairy.
   - In the name of increased milk production and profit, in countries such as the USA, dairy cows are repeatedly injected with bovine growth hormone, a genetically-engineered hormone that has been shown to increase the risk of health problems like mastitis and lameness.

The cows at the pasture releases and open farm events hybridize two categories of animals: they are objectified as invisible food animals, and subjectified and anthropomorphized animals with personal names, a narrative and a history (Stewart and Cole 2009). This ambiguous position creates a need and a challenge to negotiate and draw the boundaries between the different positions of the cows to not make it normatively threatening to consume dairy products. This boundary is drawn by the neutral language used to describe the cows and the calves as they 'give' their milk to humans, and by avoiding any ethical discussion about the production and consumption of dairy.

It could even be argued that the real cows and calves are not present at the pasture releases and open farm events. The animals that are actually there are rather signifiers, with functions to entertain, educate and promote the consumption of the real cows' bodily fluids. They are kept behind bars for children to look at and pet, turned into teaching material for parents to point at, and laughed at as they run out to escape their half year long confinement. They are constructed as if they were natural resources with their only purpose in life to fulfil human desire to consume dairy, like the fieldnote excerpt below shows:

> When we get closer to the confinement where some calves have been placed there is a tree with a sign on it telling children to "Leave your dummies here for our animal babies". We hear one parent talking to his child about the animal babies saying "they need their dummies just like you do" and "they drink their milk just like human babies". He goes on to talk about the milk and how the cows give their milk to us so that we can drink it.

In the formal education system, a regulation of eating habits takes place. This can, as Pedersen (2010) argues, be understood as an effect of power that actively produces reality and rituals of truth. To extend this Foucauldian analysis, we argue that the regulation of eating habits also produces both human and animal subjectivities. The tens of thousands of Swedish school children who are taken to open farms and pasture releases to see the cows and calves, and offered food products from cows as part of their visit (they are even reminded in notes not to bring food of their own but rather taste the things that the dairy industry is offering [Stockholms Fria 2014]) are subjected to food advertising and socialization, presented as entertainment and education. For the dairy industry, it seems to be a crucial event in the school children's education. One of the interviewed dairy industry representatives describes her reaction when hearing about a school class that cancelled their visit to the farm:

Last year, there was this class from [city] that had miscalculated and that didn't have enough money to rent a bus to go here, they lacked 3000 kronor so they informed us about this the day before, and I, I felt like almost going there and getting them myself, because, now those kids weren't able to come out that year, and who knows, they might never be able to come out to the country to see the cows, and they lose this whole thing... I think it is important, to keep this connection, because, like in Australia, in Sydney where I have been a few times, everyone drinks soy milk, because, they have no connection to the dairy industry, the land is so huge and there are no Australians working in the dairy industry, so the kids, they have no connection any longer to the milk. [Dairy industry representative 2][9]

During the visits to the farms, children are taught to approach dairy as something with an animal origin, but still something that is there for human use. They are taught to conceptually distance the animals they eat from those with whom they have an emotional bond or for whom they feel ethically responsible.

Concluding Discussion

In the autumn, when the cows after a few months on pasture are locked in again, there are no school children coming to see them returned to incarceration in order to get educated about how the production of dairy is carried out. In contrast to the boisterous spectacle of the pasture release, the reconfinement of cows is invisible and silent to the public. The cows at the pasture releases are 'feel-good cows', as one dairy industry representative notes, and this is the role they play in the dairy industry marketing image; cows who are happy and content with their place in the production processes.

Ironically, the problem that the open farm events and the pasture releases claim to be addressing in these events—how the production and consumption of dairy products have become separated to the point that, as one of the interviewees puts it, "people don't know any more where their food is coming

---

9 These claims about the dairy industry and milk consumption in Australia are however incorrrect. Australian fresh milk consumption has been steadily increasing during recent years. Also, the Australian dairy industry is still typified by relatively small producers and many herds are family owned (PricewaterhouseCoopers 2011).

from"—is reinforced in these staged events. The affectively useful, feel-good cows of the pasture releases are caricatures of themselves. They are spectacular products of *zooësis* composed of a number of other hybridized entities such as the fake, bipedal calf Kalvin; a bovine 'catwalk'; and a mechanical bull. This agri-industrial performance masquerades as education in a bizarre carnivalesque mode, while producing human and animal subjectivities that fit seamlessly within the 'happy meat' and 'new carnivore' regimes driving the dairy industry's promotional machinery.

However, if the desire for human-animal intimacy and identification—a promise delivered through the pasture releases, but one that the dairy industry will never be able to fulfil—is powerful enough to draw a crowd, it could also be redirected into a potential political force for change. This redirection would require a profound unsettling of the pasture release scenario; a violent disruption of the images and emotions it skilfully produces; in short, a 'disturbing' education (Rowe 2011) confronting the manipulative hypocrisy of dairy. Study visits to animal sanctuaries may offer a contrasting educational complement to the 'edutainment' of dairy production sites, inviting children and adults to deep reflection on what relations humans and animals may form outside regimes of commodification. To extend dairy counter-education into the classroom, we follow Gunnarsson Dinker and Pedersen (in press), suggesting critical pedagogical activities such as mapping one's own school's place in the animal-industrial complex; delineating a fuller and more accurate 'way of the milk' by mapping all phases an animal individual goes through in the dairy production system from inception to slaughter; exploring the vested interests and expansion strategies of the dairy industry that connect animal, human, and environmental exploitation; and comparing the dairy industry promotional narratives with the materials provided by animal rights organizations.

Most of all, we would like to see battalions of guerrilla teachers co-opting the pasture release and open farm events, turning them into sites of critical counter-education by exposing the inner logic of the dairy industry system as well as the histories of individual animals exploited therein.

References

Adams, Carol J. *The Sexual Politics of Meat: A Feminist-Vegetarian Critical Theory.* 20th anniversary ed. New York: Continuum, 2010.

Arla. *Kor som mår bra*, 2012. Accessed February 17, 2015. http://www.arla.se/om-arla/korna/kossan/.

Arla news. *10 år med Arlas kosläpp—intresset är större än någonsin*, 2014. Accessed November 15, 2014. http://nyheter.arla.se/2014/04/29/10-ar-med-arlas-koslapp-intresset-ar-storre-an-nagonsin/.

Arla minior. *Boka besök*, 2011a. Accessed November 15, 2014. http://arlaminior.se/for-lararen/boka-besok.

———, *Vad är Arla minior?*, 2011b. Accessed November 15, 2014. http://arlaminior.se/for-lararen/vad-ar-arla-minior.

Chaudhuri, Una. "(De)Facing the Animals: Zooësis and Performance." *TDR: The Drama Review* 51.1 (2007a) (T193): 8–20.

———, *Zooësis: Animal Acts for Changing Times*, 2007b. http://cas.nyu.edu/object/ug.academicprograms.collegiatefall2007.

Cole, Matthew. "From 'Animal Machines' to 'Happy Meat'? Foucault's Ideas of Disciplinary and Pastoral Power Applied to 'Animal-Centred' Welfare Discourse." *Animals* 1 (2011): 83–101. http://www.mdpi.com/2076-2615/1/1/83/pdf.

Crist, Eileen. *Images of Animals: Anthropomorphism and Animal Mind*. Philadelphia: Temple University Press, 2000.

Dairy Sweden *Milk. Key Figures Sweden*. LRF Dairy Sweden, 2014.

Debord, Guy. *Society of the Spectacle*. Detroit: Black & Red, 1983.

Desmond, Jane C. *Staging Tourism: Bodies on Display from Waikiki to Sea World*. Chicago and London: The University of Chicago Press, 1999.

Djurens rätt. *Mjölkfabriken—en rapport om djuren i mjölkindustrin*. Älvsjö: Djurens rätt, 2013.

Ekoweb Sverige. *Ekologisk livsmedelsmarknad. Rapport om den ekologiska branschen sammanställd av Ekoweb.nu*. Lidköping: Ekoweb Sverige, 2013.

Gillespie, Kathryn. "How Happy is Your Meat?: Confronting (Dis)connectedness in the 'Alternative' Meat Industry." *The Brock Review* 12.1 (2011): 100–128.

Gunnarsson Dinker, K. & Pedersen, Helena. (in press.) "Critical Animal Pedagogies: Re-learning our Relations with Animal Others." In *Palgrave Handbook of Alternative Education*, edited by H. Lees & N. Noddings. Palgrave Macmillan.

Gutjahr, J. "The Reintegration of Animals and Slaughter into Discourses of Meat Eating." In *The Ethics of Consumption. The Citizen, the Market and the Law*, edited by Helena Röcklinsberg & Per Sandin, 379–385. Wageningen: Wageningen Academic Publishers, 2013.

Hochschild, Arlie Russell. *The Managed Heart*. Berkeley: University of California Press, 1983.

Holloway, Lewis. "Subjecting Cows to Robots: Farming Technologies and the Making of Animal Subjects." *Environment and Planning D: Society and Space* 25 (2007): 1041–60.

Jönsson, Håkan. *Mjölk: en kulturanalys av mejeridiskens nya ekonomi*. Eslöv: Symposion, 2005.

Lidfors L., Stehulova, I. & Spinka, M. *Ko-kalvseparation: Mindre stress när mjölkkor och kalvar skiljs tidigt*. SLU, Fakta Jordbruk 13, 2004.

Molloy, Claire. *Popular Media and Animals*. Houndmills, Basingstoke, Hampshire: Palgrave Macmillan, 2011.

Parry, Jovian. "Gender and Slaughter in Popular Gastronomy." *Feminism & Psychology* 20.3 (2010): 381–396.

Pedersen, Helena. *Animals in Schools: Processes and Strategies in Human-Animal Education*. West Lafayette, Ind: Purdue University Press, 2010.

Potts, Annie, Armstrong, Philip, and Deidre Brown. *A New Zealand Book of Beasts. Animals in our Culture, History and Everyday Life*. Auckland: Auckland University Press, 2013.

PricewaterhouseCoopers. *The Australian Dairy Industry. From Family Farm to International Markets*. PricewaterhouseCoopers International, 2011.

Rowe, Bradley D. "Understanding Animals-Becoming-Meat: Embracing a Disturbing Education." *Critical Education* 2.7 (2011). Accessed September 28, 2014. http://ojs.library.ubc.ca/index.php/criticaled/article/view/182311.

Röcklinsberg, Helena, and Sandin, Per. (eds.) *The Ethics of Consumption. The Citizen, the Market and the Law*. Wageningen: Wageningen Academic Publishers, 2013.

Skånemejerier. *Vill du besöka en gård?*, 2010. Accessed November 15, 2014. http://www.skanemejerier.se/sv/Mjolkgardar/Vill-du-besoka-en-gard/.

———, *Skånemejeriers kosläpp 2014*, 2013. Accessed November 15, 2014. http://www.skanemejerier.se/sv/Mjolkgardar/Besok-ett-av-vara-koslapp/?utm_source=apsis-anp3&utm_medium=email&utm_content=unspecified&utm_campaign=unspecified.

Sorenson, J. (ed.) *Thinking the Unthinkable: New Readings in Critical Animal Studies*. Toronto: Canadian Scholars' Press, 2014.

Stănescu, Vasile. "Crocodile Tears, Compassionate Carnivores and the Marketing of 'Happy Meat.'" In *Thinking the Unthinkable: New Readings in Critical Animal Studies*, edited by John Sorenson, 216–233. Toronto: Canadian Scholars' Press, 2014.

Statistics Sweden. *Yearbook of Agricultural Statistics 2013 Including Food Statistics*. Örebro: Statistics Sweden, Agriculture Statistics Unit, 2013.

Stewart, Kate, and Cole, Matthew. "The Conceptual Separation of Food and Animals in Childhood." *Food, Culture & Society* 12.4 (2009): 458–476.

Stockholms Fria. *Arla tonar ner reklam till barn*, 2014. Accessed November 15, 2014. http://www.stockholmsfria.se/artikel/114503.

Svenskt Kött. *Om kött: uppfödning*, 2014. Accessed February 17, 2015. http://www.svensktkott.se/om-kott/uppfodning/.

Swedish Board of Agriculture. *Market Overview—Milk and Dairy Products* Jönköping: Swedish Board of Agriculture, Division for Trade and Markets, 2012.

———, *När dina nötkreatur är ute eller på bête*, 2014. Accessed February 17, 2015. http://www.jordbruksverket.se/amnesomraden/djur/olikaslagsdjur/notkreatur/utevistelseochbetesgang.4.4b00b7db11efe58e66b8000308.html.

Sydsvenskan. *Detta gillar skåningarna*, 2012. Accessed November 2, 2014. http://www.sydsvenskan.se/ekonomi/detta-gillar-skaningarna1.

Thierman, Stephen. Apparatuses of Animality: Foucault Goes to a Slaughterhouse. *Foucault Studies* 9 (2010): 89–110.

Vialles, Noëlie. *Animal to Edible*. Cambridge: Cambridge University Press, 1994.

CHAPTER 7

"Peace and Quiet and Open Air": *The Old Cow Project*

Melissa Boyde

If there is any truth to a claim made by organizers of a campaign against the slaughter of cattle, reported in the newspaper *India Today* (1996), that "the trembling and wailing of the cows being slaughtered lead to earthquakes" then the planet is in for a rough ride (Fox 1999, 27)

∴

Snapshots

After the catastrophic 2011 Japanese earthquake and tsunami an exhibition titled *Lost and Found Project* was curated from the roughly 750,000 photographs found scattered through the mud and debris.[1] The photographs, snapshots of life, were mostly damaged and the images of the subjects were hard to discern, merely traces. Two months after the earthquake a group formed to sort the photos; calling themselves the "Memory Salvage" project they cleaned, numbered, digitized and indexed each of the photographs.

In an essay for the exhibition art historian Geoffrey Batchen writes:

> Like every photograph, the snapshot is an indexical trace of the presence of its subject, a trace that both confirms the reality of existence and remembers it, potentially surviving as a fragile talisman of that existence even after its subject has passed on. (Batchen)

On a poster that accompanies the exhibition are a series of questions: "What are we supposed to feel and think when we look at these pictures? Should we be happy they were found at all, or sad that they will never be returned to their owners? Or should we simply mourn for the dead? The more I struggle to find

---

[1] I am grateful to my colleague Anne Collett for bringing this exhibition to my attention.

FIGURE 7.1 *Lost and found.*
PERMISSION TO REPRODUCE GRANTED BY MUNEMASA TAKAHASHI DIRECTOR, *LOST AND FOUND PROJECT*.

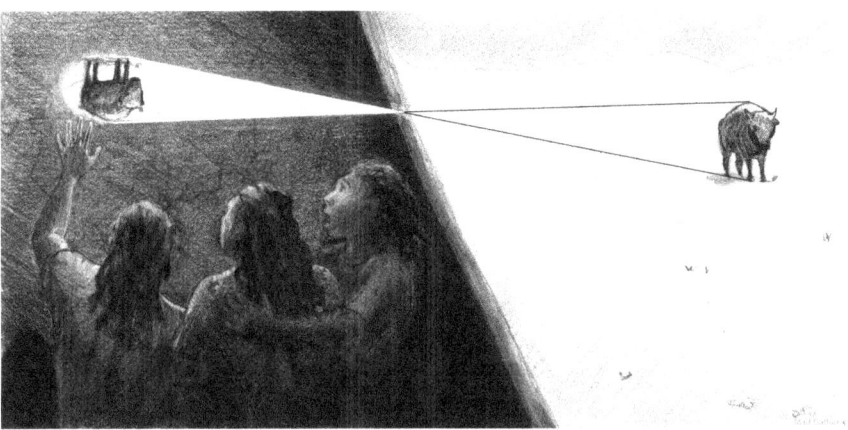

FIGURE 7.2 *Paleo-camera.*
©MATT GATTON.

answers, the more missing pieces I seem to find. But without looking at the pictures, I do not think we will see anything at all".

The pictures in this chapter form part of what I call *The Old Cow Project*. A number of them are of a particular herd of cattle which had its beginnings almost thirty years ago on a small acreage I had just moved to on the outskirts of Sydney. A recent series of photographs taken of the herd by artist Derek Kreckler shows them at their current home, on coastal pastures south of Sydney. In contrast to the lives of the majority of cattle in Australia who are generally regarded as livestock in the cattle industries—primarily bred for either beef or dairy products—the members of this herd live out their life cycle. The snapshots from my photo albums included in this chapter, and the artist's photographs, gesture to the lives of cattle obscured in representation, lives untraced, from the early *camera obscura* to the present.

The *camera obscura*, the chamber of darkness, typically has a darkened enclosed interior space, box-like in shape, with a small hole through which light passes. An inverted image of an exterior view is produced within the chamber, reduced in size but in exact proportion to the original and with the illusion of depth, color and movement. Jonathon Crary points out that for over two hundred years the *camera obscura* "stood as a sovereign metaphor...of how observation leads to truthful inferences about the world"; it was "a site at which a discursive formation intersects with material practices" (1993, 31). Like other technologies of seeing, the *camera obscura* is "not ideologically neutral" and "implies not only the existence of the 'real' but a specific relation to it: a

FIGURE 7.3 *Snapshots 1989–2015.*

given which is always projected in inversion" (Camera Obscura Collective 2010, 272–73).

This chapter suggests that in this topsy-turvy world, in which the 'real' rendered through representation is based on inversion, it seems predictable, if not paradoxical, that humans (never ideologically neutral) render the non-human as inversion—in the sense felt by Others of whatever kind. Therefore, mindful of the convolutions and limits of representation whereby "the real is read through representation, and representation is read through the real" (Phelan 1993, 2), *The Old Cow Project* incorporates photographs to show those whose lives are culturally overlooked or deemed unimportant and who, relegated to the margins, remain untraceable (except perhaps in memory or imagination). First are some snapshots from my photo albums. Minnie and Moo were the original members of the herd brought together by two separate purchases I made at the local saleyard about 25 years ago. I had only recently moved to the country from the inner city and thought I needed a cow to eat the grass. I knew nothing about cows. I found the local saleyard, decided on a black and white calf in a yard at the back, and waited some hours until she was herded into the yard where I bid for her against a bevy of cattlemen and butchers. I arranged for her to be brought home on a truck and called her Moo. But Moo was unsettled, always mooing and searching, so I went back to the next sale and bought Minnie, just a couple of months old. They were immediately inseparable and likewise I instantly fell in love with them both, and later with their offspring and with each of the other cows and steers who one way or another became part of their herd. It has been quite a few years now since Minnie and Moo died.

> Memory. Sight. Love. It must involve a full seeing of the Other's absence (the ambitious part), a seeing which also entails the acknowledgment of the Other's presence (the humbling part). For to acknowledge the Other's (always partial) presence is to acknowledge one's own (always partial) absence. (Phelan 1993, 149)

Memorials

About 20 years ago, en route to somewhere, I stopped at a park in Homebush Bay in Sydney for some refreshment and found a small café set amidst gardens. What I didn't know, until I came upon a plaque placed on the side of a garden bed, was that I was on the former site of an abattoir. From memory the plaque stated something like: "In memory of the many animals who died at the State Abattoir". I was surprised at first and then pleased that the memorial plaque was there. But at the same time I felt somehow disturbed by it. When I went back recently to photograph the plaque it was gone and all my enquiries came to nothing.

The State Abattoir at Homebush was established on that site one hundred years ago. Prior to this, from 1860, animals were slaughtered at an abattoir on Glebe Island, a few kilometers from the centre of Sydney. Blood and offal were dumped in the surrounding harbor waters, described as often being "blood red" (Glebe Society Bulletin 2006, 5). In close vicinity to the abattoir various associated commercial industries were rife such as a tannery, fertilizer factory and candle and soap makers. By the turn of the century the abattoir buildings were considered to be "of wrong design, inadequate to the business which should be done in connection with them" and overall were "hopelessly out of date" (Rosen et al., 1990, 18). The growth of the residential areas and "alarm at the outbreak of plague in Sydney" contributed to the appointment in 1900 of a Parliamentary Standing Committee to assess the possible relocation of the abattoir to another site (Rosen et al., 1990, 18). The report outlined the "need to eliminate or reduce the cattle driving nuisance occasioned by stock being driven to slaughter through commercial and residential areas" (Rosen et al., 1990, 18–20). The more rural site at Homebush, although still reasonably close to the city, was perfect for this out of sight, out of mind requirement.

When it opened for business the new State Abattoir at Homebush was the largest abattoir in Australia (Sydney Olympic Park Authority "Conservation Management Plan" 2003, 15). By the end of that century it had been shut down and later demolished to make way for the 2000 Sydney Olympic Games stadium and associated hotels and other buildings. All that now remains of the abattoir is the administration precinct, an impressive group of buildings "designed in Federation (Arts and Craft) style...set amongst landscaped gardens and lawns" (Sydney Olympic Park Authority "Conservation Management Plan" 2003, 15–16). The gardens were designed "to provide a calm and restful environment for the Abattoir administrators" (Ibid, 16). The area that encompasses the surviving buildings and gardens is now called the "Abattoir Heritage Precinct", developed as part of the Sydney Olympic Park Authority's (SOPA) Conservation Management Plan. A note about the gardens in the plan, states that "the rose garden provided roses for female visitors at lunch" (Sydney Olympic Park Authority "Conservation Management Plan" 2003, 42). This odd statement, which indicates that luncheons were held in the abattoir precinct, not only reveals gendered behaviors but also the apparent ability to act out regimes of 'civility' while large-scale sanctioned slaughter was taking place on the other side of the hedges which surround the gardens.

The animals are now long gone but the rose garden remains, carefully tended. Amidst the plots with various rose specimens are art installations, created as part of the Sydney Olympic Park public art commissions. Among the art objects are several wooden park benches with outlines of the head of a bull, a sheep and a pig carved into the wooden backs.

At one edge of the garden, on "Meatworks Path", lies the "Australia Map Mosaic" which depicts the heads of a pig, a sheep and a steer over a map of Australia, with an inscription underneath which gives the dates of the operation of the abattoir (Sydney Olympic Park Authority "Public Art", 12).

In an opposite corner of the garden (across the road from what was once the terminus for the cattle trains which brought the animals from near and far to be sold and killed but which is now the main railway station for Sydney Olympic Park) is an art installation which "attempts to recreate the atmosphere" of the abattoir during its 74 years of operation (Sydney Olympic Park Authority "Public Art", 13). The work includes five animal drinking troughs salvaged from the abattoir. The troughs are buried in the ground and laid out to resemble graves.

At the head of each trough is a timber 'headstone' with an illustrated ceramic tile attached, each featuring a line from a nursery rhyme:

FIGURE 7.4 *Carved timber bench, Sydney Olympic Park Urban Art Collection.*

FIGURE 7.5 *Australia map mosaic, Sydney Olympic Park Urban Art Collection.*

'Clickety Click Clickety Clack, Hear the wheels of the railway track'
'To market to market, Jiggety Jig Jiggety Jog'
'Here a Moo, there a moo, everywhere a moo, e i e i o'
'Flipperty Flop Flipperty Flop here comes the butcher to bring us a ....'
'Little boy blue come blow your horn'

From the start the State Abattoir was flawed and inadequate. When the first stage of the building works was completed in 1912 there were numerous problems, including that facilities for treating offal were non-existent (Rosen et al., 1990, 8). Modifications were made—"blood was to be disposed of via a gravitation system; the need for killing chambers to be constructed prior to the commencement of killing was recognised", and so on (Rosen et al., 1990, 8). Building work continued, and on April 7, 1915 the abattoir was officially opened by the then Premier of NSW. It was a grand affair—guests were transported there in special trains which departed from Central Station in Sydney. The carcass of the first sheep killed was presented to the Premier and later sold at an auction in aid of the Belgian wartime fund (Rosen et al., 1990, 24).

FIGURE 7.6 *Drinking trough animal memorial, Sydney Olympic Park Urban Art Collection.*

FIGURE 7.7 *Nursery rhyme ceramic headstone tiles, Sydney Olympic Park Urban Art Collection.*

Despite the many deficiencies of the building by 1923 it was considered to be "Australia's largest and most modern abattoir" and up to 1600 men were employed there to manage and conduct the slaughter of up to 18,000–20,000 sheep, 1,900 cattle, 2,000 pigs and 1,300 calves per day (Rosen et al., 1990, 12).

By the 1940s there was significant deterioration of the plant and equipment and funds were only allocated to address the most urgent problems. In some cases "second hand equipment was purchased to replace equipment beyond repair" (Rosen et al., 1990, 14). However a documentary film about the State Abattoir, *Meat… with care!* shows a different story. Premiered at the lavishly ornate State Theatre in Sydney, the documentary style film promotes the concept of a modern facility producing clean, disease-free meat. No actual killing of the animals is shown. Instead abattoir workers are depicted during recreation, playing sport or chatting in the men's locker room. The Grand Parade of animals at the Sydney Royal Easter Show, a crowd pleaser directed at families, is also shown. The State government Meat Industry Board which funded the film would no doubt have been satisfied that by August the year of its release *Meat… with care!* was on the program of the Edinburgh Film Festival.[2]

Four years later, from 1965 to 1976, $27 million was spent upgrading the facilities at the abattoir but the inherent flaws in design and construction meant that much of the money was spent covering up structural problems

---

2 My thanks to Marc David Jacobs at the Edinburgh International Film Festival (EIFF) for providing information on the 1961 programme which lists the screening time as Tuesday 29th August at 6.15 p.m.; and for the following information: "EIFF was very much in transition during the early 1960s, as it grew out of its origins in screening a great deal of industrial documentaries (including a regular amount of content from the Australian National Film Board). So the 1961 Festival would have been a bit of a mix of educational/specialized 16mm documentaries and Eastern European features, alongside films by Ingmar Bergman and Chris Marker" (pers. comm.by email 2/6/2014).

FIGURE 7.8 *Meat... with care!*
REPRODUCED COURTESY OF STATE RECORDS NSW: NRS 5194, POSTERS AND BROADSHEETS, HOMEBUSH ABATTOIR CORPORATION; POSTER, 'MEAT WITH CARE', 1961.

(Sydney Olympic Park Authority "Industrial History"). For example, in order to "hide the decay from the Americans reviewing the facilities to assess their compliance ... for the issue of an export license", crumbling internal walls were covered over with stainless steel panels (Rosen et al., 1990, 31). To further complicate matters "the upgrade appeared to be badly directed and wasted as the management appeared to lack clear direction and forward planning" (Sydney Olympic Park Authority "Industrial History"). Despite all of the money spent the abattoir was finally shut down in 1988. All of the buildings involved in the processes of killing animals have since been demolished: the Saleyards (now the site of the Sydney Olympic stadium), the Pork, Veal and Mutton Killing and Dressing areas; the Mutton Grading Building, which included the sheep "Head Splitter"; the "Beef Slaughter Hall", including the "Johnson Upward Hide Puller"; the Meat Chilling Complex, including the Boiler House, Cannery and Blast Freezer; the By-Products Treatment Complex, and the Meat Market buildings where the slaughtered animals were sold as meat.

The only material remains of this multi-million dollar building in which literally millions of animals died are the five buried water troughs, which form part of the memorial in the administration building gardens. Alfred Loos famously argued that the purpose of architectural monuments is ethical—the external marker prompts an internal reckoning: "a confrontation with death prevents us from going on with the usual business of life ... it carries us to another place, a place, usually submerged, within the self" (Bonder 2009, 63). Does the row of the five buried water troughs with their nursery rhyme headstones confront us with the mass animal death that took place on that site and somehow take us to another place submerged within the self? Do the monuments in the abattoir heritage garden create what some suggest is the role of public memorials "to create a theater for memory capable of embodying truths that make it possible to affirm life and contemplate a better future"? (Bonder 2009, 65). Memorials produce meaning as a reminder of the past in the present but what if the events that took place in the past remain as a continuous present, with the number of animals killed in slaughter factories reaching new heights? And what if this mass slaughter of animals is (deliberately) culturally obscured?

The Trace

If there is little visible evidence of the lives of the animals caught in the livestock industries in Australia's past, or in the present, other than temporary traces of their carved up remains—refrigerated, transported, displayed for purchase, cooked, eaten, discarded ... If nothing of the fleshed, living, breathing body remains... If there is no (non-human) language discernible to humans to tell any of their stories... If, as the poet writes:

> Nobody
> bears witness for the
> witness. (Celan 2001, 258)

All that remains to render some of what happened to animals inside the State Abattoir are traces in the official archive. A series of photographs that form part of the archival documents record the operations of the abattoir during the 1930s and 40s. Always the 'witness' of something, the archive "demands to be constructed" (Farge quoted in Didi-Huberman 2008, 99). Every trace and "each discovery emerges from it like a *breach in the history conceived*, a provisionally indescribable singularity that the researcher will attempt to weave into the fabric of everything he or she already knows, in order to produce, if possible, a *rethought history...*" (Didi-Huberman 2008, 99).

Searching for traces—I returned to Homebush and went into the new café in the Heritage Precinct, not for a snack this time but to ask if anyone on the staff there had seen the small memorial plaque I had stumbled upon some years earlier; but nobody had. Blues music was softly playing, the atmosphere was inviting, relaxing. At the back of the café, hanging from stainless steel hooks were salamis and other cured meats. On the way out I noticed the name above the entry door—Abattoir Blues café.

> The trace is
> a hold/a hole
> of evanescence through which
> travel small powerful things,
> impotently, earnestly, but, and,
> whether, what if underlying them.
> Traces of what happened
> commend your attentiveness to the almost invisible.
> Since they cannot command.
>
> For once hosed down,
> pavement no longer
> shows its bloody brown. (Du Plessis 2013, 25)

Phelan writes of photography's power as ubiquitous witness to the traumas of history and our own capacity to play a part in their unfolding:

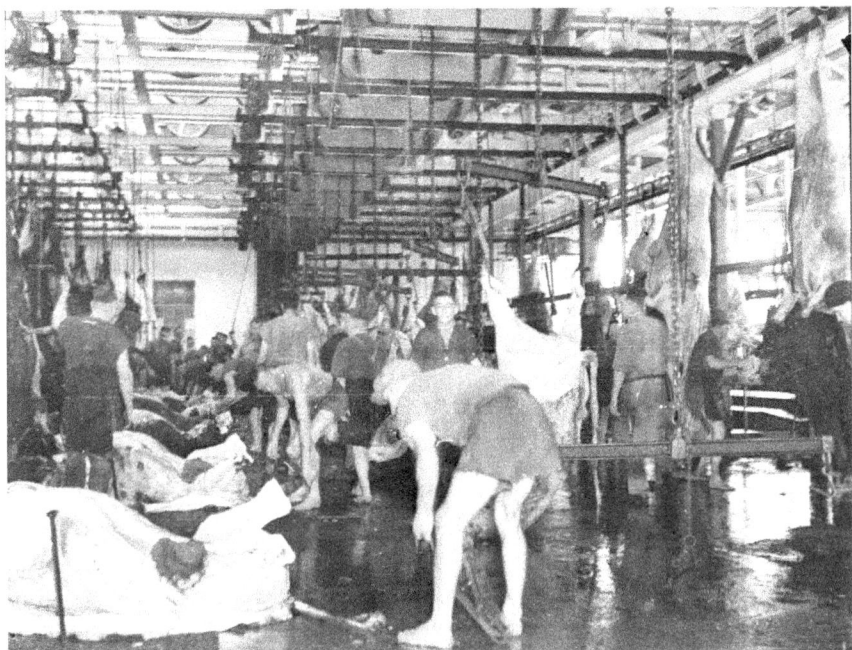

FIGURE 7.9 *Homebush Abattoir 1937.*
COLLECTION OF THE STATE LIBRARY OF NSW.

"photography inhabits two temporal registers—the now staged in the moment of its making and the now of the moment it is seen" (Phelan 2010, 51). I re-present this photo from the archives as a trace of the trauma for the animals killed at the State Abattoir and as an observation of what is not seen in the now of the moment as the majority of us turn a blind eye to "our own capacity to play a part" in the unfolding of events.

To remember and bring into the present the lives of the millions of animals used in the livestock industries and killed at abattoirs scattered throughout the country—most of whom are young, some just babies, some older dairy cows who no longer produce enough milk to meet the set quotas, and others who are sick or injured—I present photographic portraits of the herd in my snapshots by artist Derek Kreckler.

In a statement about his cow series Kreckler writes:

> Contrary to the romantic perception of a sleepy beast grazing on pastures green and brown the process of life and death has become efficient through mechanical and brutal processes. The ritual of death has become a banal science out of the sight and mind of the consumer.

> When the cycle of breeding for abattoir is broken we are able to witness cows removed from a mechanistic cycle of life and death.
>
> It was a privilege for me to capture a very brief moment in the lives of the cattle in this herd, likely some of the oldest, if not the oldest, cows in Australia. In this case the noun *herd* is probably a misnomer because these animals have personalities, likes and dislikes and seem to gather for collective reasons rather than because of an imposed process. Being with them gave me pause for thought.

Following Jean-Luc Godard's suggestion that "montage... is what makes visible" (quoted in Didi-Huberman 2008, 138) I note that just down the road from where the herd in Kreckler's photographs lives is the local cattle saleyard. On the same day that the artist took these photographs other calves and cattle were penned in holding yards at the sale, surrounded by bidders. Many one-week-old calves were there for sale, as they are most weeks. Deemed to be a by-product of milk production the calves are taken from their mothers within twelve hours of their birth, as per the recommendation of the dairy industry (The Life of the Dairy Cow, 2015, 21).

If given the opportunity to be together, a close, loving, lifelong relationship would most likely have evolved between calf and cow. The most senior cow in the herd at our farm, and perhaps now the oldest cow in Australia, had a calf seventeen years ago who we named Snowy. Until Snowy's recent death she and her mother had always been inseparable. During Snowy's protracted illness her mother sat with her, always grazed within her view and every day licked her all over—just as she had done when Snowy was born. The visible bond between them was undeniable.

At the saleyard, one by one, sometimes as a 'job lot', the calves went under the auctioneer's hammer. A large covered-in cattle truck later arrived to transport them to a slaughter factory.

Phelan writes "that by seeing the blind spot within the visible real we might see a way to redesign the representational real" (1993, 3). Returning to the *camera obscura*, the darkened enclosed interior space with inverted (upside down, back to front) moving full color images, for centuries accepted as an observation of "truthful inferences about the world" (Crary, 1993, 29). In my mind I see the *camera obscura* superimposed onto the culturally closed off space of the abattoir with its box-like, high walls built to block out exterior vision.

In the stillness of the darkened space, lit only by light from a small high window, the inverted view is not the trace of an exterior reality—if that were the case the image might be of the charming rose gardens, with perhaps a ghostly

"PEACE AND QUIET AND OPEN AIR"   143

FIGURE 7.10A   *Milo, Bonnie, Star, Sweetie and Billie.*

FIGURE 7.10B   *Benson and Murray with Milo in the background.*

FIGURE 7.10C   *Minuet, Snowy and Mintie.*
OLD COW PROJECT SERIES. ARTIST DEREK KRECKLER.

FIGURE 7.11   *Homebush Abattoir 1931.*
COLLECTION OF THE STATE LIBRARY OF NSW.

glimpse of roses being cut for the ladies who lunched with the administrators of the abattoir. The view instead is of the remains of animals to whom we turn a blind eye. Officially marked as "graphic material", the real of representation are the partial bodies of animals, hanging upside down and lifeless on metal hooks in the Chilling Halls.

The interior real of the slaughter factory hidden from view to all except the witnesses (the animals and the paid employees) is partially captured in images from the government archive which provide a trace of the untimely deaths of the multitudes of animals strung up and killed every day.

These archival 'traces' hold the possibility of awakened memory and "rethought histories" (Didi-Huberman 2008, 99). But the very same photographic traces, showing the dismembered bodies of animals led to slaughter at the State Abattoir, serve to keep each cow less-than-whole; (literally) rendering them as unlovable and thereby mitigating the possibility of love and by extension, prohibiting ethical treatment.[3]

---

3 I am grateful to Peggy Phelan for her engagement with these ideas.

FIGURE 7.12   *Beef carcasses 1954.*
COLLECTION OF THE STATE LIBRARY OF NSW.

It should be enough to make the earth tremble.

From the debris of the catastrophic effects of the Japanese tsunami emerge the photographic traces of someone's loved ones, damaged and almost unrecognizable. The *Lost and Found Project* exhibition poster suggests "without looking at the pictures, I don't think we'll see anything at all".

FIGURE 7.13   Old Cow Project *series. Artist Derek Kreckler.*

From the debris of the catastrophic effects of the slaughter factory emerge barely a trace of other cows' loved ones—each body is damaged and unrecognizable. Thus represented, the photographic 'trace' of a history of animal slaughter (already hidden and obscured) makes the life gestured to in the artist's photographic portraits of (be)loved cows 'unthinkable' and literally un-imaginable.

Mindful of the (deliberate) inversions of representation, and therefore without any certainty... *The Old Cow Project* nevertheless submits that: "without looking at the pictures, I don't think we'll see anything at all".

References

Batchen, Geoffrey. "Lost and Found Project: Family Photos Swept by 3.11 East Japan Tsunami." Catalogue essay. http://lostandfound311.jp/en/essay/index.html.

Bonder, Julian. "On Memory, Trauma, Public Space, Monuments, and Memorials." *Places* 21.1 (Spring 2009): 62–69. http://www.escholarship.org/uc/item/4g8812kv.

Camera Obscura Collective. "Feminism and Film: Critical Approaches." In *The Feminism and Visual Culture Reader*, edited by Amelia Jones, 268–72. 2nd ed. New York: Routledge, 2010.

Celan, Paul. "In Prague." *Selected Poems and Prose of Paul Celan*, edited by John Felstiner, 258. New York: W.W. Norton 2001.

Crary, Jonathon. *Techniques of the Observer: On Vision and Modernity in the Nineteenth Century*. Cambridge, MA: MIT Press, 1993.

Didi-Huberman, Georges. *Images in Spite of All: Four Photographs from Auschwitz*. Translated by Shane B. Lillis. Chicago: University of Chicago Press, 2008.

Du Plessis, Rachel Blau. "Draft 87: Trace Elements." In *Poetry & the Trace*, edited by John Hawke and Ann Vickery. Glebe, NSW: Puncher and Wattman, 2013.

Fox, Michael W. "India's Sacred Cow: Her Plight and Future." *Animal Issues* 3.2 (1999): 27.

Glebe Society Bulletin 3 (2006). http://bulletin.glebesociety.org.au/2006_03.pdf.

Kreckler, Derek. "The Old Cow Project." Artist's statement, 2013.

"Lost and Found Project: Family Photos Swept Away by 3.11 East Japan Tsunami." http://lostandfound311.jp/en/.

Meat... with care! Australian Instructional Films. Prod., Dir: Lex Halliday. Script: Alan Seymour and Lex Halliday. Sound: Don Connaly. Commentators: Frank Waters and James Condon. 20 minutes.

Phelan, Peggy. *Unmarked: The Politics of Performance*. London: Routledge, 1993.

Phelan, Peggy. "Haunted Stages: Performance and the Photographic Effect." In *Haunted: Contemporary Photography, Video, Performance*, edited by Jennifer Blessing. New York: Guggenheim Museum Publications, 2010.

Rosen, Sue, Irving, B., Godden, D., Mitchell, S., and Tzavaras, P. "State Abattoir, Homebush. A History and Record of the Buildings, Structures and Technology." 1990. http://dx.doi.org/10.4227/11/50B40D0EA5E71.

Sydney Olympic Park Authority. "Conservation Management Plan: Abattoir Heritage Precinct." Graham Brooks and Associates, June 2003.

Sydney Olympic Park Authority. "Industrial History." NSW Government. http://www.sopa.nsw.gov.au/our_park/history_and_heritage/industrial_history.

Sydney Olympic Park Authority. "Public Art." NSW Government. http://www.sopa.nsw.gov.au/data/assets/pdf_file/0007/949471/Fact_Sheet_-_Public_Art.pdf.

"The Life of the Dairy Cow: A Report on the Australian Dairy Industry." Sydney: Voiceless: The Animal Protection Institute, 2015. https://www.voiceless.org.au/sites/default/files/The%20Life%20of%20the%20Dairy%20Cow.pdf.

CHAPTER 8

"Do You Know Where the Light Is?" Factory Farming and Industrial Slaughter in Michel Faber's *Under the Skin*

Kirsty Dunn

> Surely meat is the most magic and beguiling of commodities—one that just *appears*, bearing almost no trace of its brutal origin. That rows of wrapped, severed cubes of flesh, perhaps adorned with labels decorated with cartoon pigs or cows, are *just there in the shop* [...] is one of the strangest normal things in our world. Meat [...] can be defined as the paradigmatic commodity...
>
> CHRIS OTTER, "Civilizing Slaughter: The Development of the British Public Abattoir, 1850–1910" in *Meat, Modernity and the Rise of the Slaughterhouse* (2008, 106)

∴

> Nestled in greaseproof paper, it winked at her, still moist and warm, irresistible and disgusting at the same time. She'd eaten it, even licked the juices from the creases in the paper, but she never mentioned it to Ensel afterwards, and that was the end of that.
>
> MICHEL FABER, *Under the Skin* (2000, 91)

∴

Despite the fact that most of the meat available in supermarkets and fast food outlets now comes from animals raised in intensive 'factory farm' conditions,[1] many Western consumers remain 'in the dark' about what exactly goes on behind the closed doors of these operations; what *does* go on is a far cry from

---

1 According to CIWF (Compassion in World Farming), approximately two of every three farm animals are raised in these intensive operations, which equates to over 50 billion animals each year. See http://www.ciwf.org.uk/factory-farming/.

nostalgic notions of the free-range animals raised on Old MacDonald's farm. Fungal infections and parasite infestation (Tietz 2010, 111), sickness-induced cannibalism (Ibid 110), lack of sunlight (Ibid 110), teeth cutting, tail docking (Scully 2010), debeaking, and suffocation (Johnson "Factory Farming" 30), are just some of the realities of factory farming. One author describes his visit to a mass-confinement hog farm as a place where straw-deprived hogs chew on bars and chains or "just lie there like broken beings" (Scully 2011, 23); "the pigs remain in a state of dying until they're slaughtered", writes another (Tietz 2010, 111). The conditions for factory farmed broiler chickens are much the same. Up to thirty percent of chickens suffer "severe lameness and swelling" and days spent standing on their own fecal matter causes breast blisters and foot-pad dermatitis (Potts 2012, 155). In Britain alone approximately 45 million chicks die before reaching "market weight" (155) due to the harsh conditions of these intensive operations. That these realities remain largely unexposed is due in part to the hidden location of these 'meat-factories', their closed and impenetrable design, and the strict enforcement of laws and regulations which prevent the entry into or recording of such operations. These factors, coupled with euphemistic language or "double-speak" (Glenn 2004, 68) and imagery that still totes these facilities as 'farms' and dying animals as "vulnerable production units" (Imhoff 2010, 74) ensures that welfare of the animals inside these factories are by in large ignored by those demanding cheap meat products.

The violent process of slaughter is also an aspect of meat production that is ignored by many Western consumers; "That animals are killed for humans to eat meat is obvious to the point of banality...." writes sociologist Nick Fiddes, "... however, the inherent conquest is rarely discussed overtly in the context of food provision" (1991, 44). In the same way that factory farming, and the unstable boundary between the 'human' and the 'animal' are largely omitted from the Western carnist conversation, the slaughterhouse too, despite its integral function, is often left uninterrogated and the reason for that omission is one and the same: it is difficult to critique something that is perpetually in hiding. "To enable us to eat meat without the killers of the killing, without even—insofar as the smell, the manure, and the other components of organic life are concerned—the animals themselves", notes critic Timothy Pachirat "this is the logic that maps contemporary industrialized slaughterhouses" (2011, 3). The violent realities of industrial slaughter must be hidden to 'legitimize' and perpetuate animal killing on the exponential scale that is needed to satisfy profiteers and meet the public demand for meat and other animal products (Elder 1998, 197). Consequently, when we do venture inside the factory farm or slaughterhouse, we usually must enter "through the writings of someone else"

(Adams 2010, 51), and Michel Faber's *Under the Skin* (2000) provides us with that opportunity.[2]

In this novel Faber replaces the human hunter, worker, and consumer with an extra-terrestrial species; and the consumable 'animal' with humans themselves—a literary device that illuminates many hidden and problematic aspects of Western meat production and consumption. By pulling back the curtain on the factory farm and 'disassembly' line, and revealing the ways in which physical and linguistic distance serve to hide these aspects of meat production, Faber exposes some harsh realities that for so long have remained 'off the table' for discussion.[3]

The protagonist of *Under the Skin* is Isserley, a female extra-terrestrial and hunter of human ('vodsel') hitchhikers.[4] In order to blend in with the vodsel population and entice her prey into accepting a ride, she has had numerous surgical procedures to alter her appearance. Instead of walking on all fours like the rest of her kind, she has become bipedal which causes great pain and discomfort. She also regularly shaves her body to remove the thick hair of her species, laments the large breasts that have been added to her frame, and refrains from clenching her fists because of the pain caused by the removal of her sixth fingers. The narrative initially follows Isserley's 'hunting' process and the uneasy conversations between herself and her targets, and then slowly reveals the fate of her captives. Stripped of their tongues, hair, and genitalia, they are held underground in factory farm-like conditions before being processed for meat and transported back as delicacies to the tables of the wealthy

---

2 The footage recorded inside these operations and shared by animal welfare organizations such as People for the Ethical Treatment of Animals (PETA) and SAFE for Animals (a New Zealand organization) are also a 'way in' for the consumer. However, it is arguably easier to read about such operations than to view them; some parts of the world carry harsh penalties for the release of such recordings and facilities are kept highly secure to prevent unauthorized visitors.

3 Faber has pointed out that *Under the Skin* is not "a tract for vegetarianism" and that he is not a vegetarian himself. (Interview with Ron Hogan, www.beatrice.com/interviews/faber/). He has, however, declared that he has strong feelings about issues concerning meat production. In an interview with Jill Adams he stated: "The trouble with our carnivorous society is that we have millions of people eating vast amounts of meat but not wanting to take moral responsibility for how it's produced. Animals can be cruelly treated and even genetically turned into monsters, as long as it all happens in secret and the result is disguised in a neat supermarket package..." (see www.barcelonareview.com/29/e_mf_int.htm).

4 In Faber's novel, 'humans' are what we would refer to as 'animal', or, more specifically, 'alien', and 'animals'—referred to as 'vodsels'—are what we would call 'humans'.

in Isserley's home land. Isserley's role as a seductive hunter, her brief encounter with Amlis Vess (vegetarian, animal rights proponent, and reluctant heir to the corporation which owns the production plant), her experience as a transgressive, hybrid 'other' figure, and her conflicted attitude towards the 'vodsels' she encounters provide compelling avenues for the critique of the modern-day meat industry and the 'disposable bodies' created by the discourses at work within this arena.

Hiding in Plain Sight: Distancing the Consumer from the Meat Production Process

The physical distance between factory farms, slaughterhouses, and the consumers that sustain them is a clear target of critique in Faber's text. Ablach Farm—the site of confinement *and* slaughter of the vodsels in the narrative—is located in back country Scotland where the roads and tracks are "deserted" (101), the farm boundary backs onto a forest, and a mountain range and sea are found close by. The placement of the facility away from general observation is a direct referral to the way Western factory farms are intentionally located far from public view. The idea that 'natural' features such as the sea, the forest, and the mountains aid in the 'hiding' of the facility is also a reference to the way nostalgic images of farms mask the realities of industrialized agriculture. "For most of us" writes critic Andrew Johnson "the word 'farm' evokes an image of a plain but dignified dwelling set in a patterned landscape of neatly tended fields; green with peacefully grazing stock, brown with clean new furrows, or golden with the rippling wealth of harvest…" (3). Thus, the natural 'barriers' to sight in Faber's novel mirror the nostalgic images of nature consumers prefer to conjure when they consider the origins of the animals they eat. In addition, the reference to "derelict cattle sheds […] eerie with emptiness, their floors moated with a slurry of rainwater and the compost of cows long gone", the "old granary" and the stable with "inanimate" contents (97–98), may allude to a decline in 'traditional' farms which may be well be considered an 'endangered species' in the increasingly industrialized agricultural landscape (Stull and Broadway 27); it is telling that these disused apparatus are also used to disguise the realities of the operations at Ablach Farm.

In addition, the 'human' (alien) owners of Ablach Farm have gone to great lengths to hide the actual enclosures where vodsels are kept.[5] They are

---

5 Of course on a practical level, the location of the vodsel pens and slaughtering facilities underground is necessary for the works to remain hidden from the vodsel public, so that the

situated on the lowest level of the subterranean plant, four levels underground. In contrast, and providing yet another nod to the lack in consumer understanding of the meat production chain, the workers' kitchen and recreation hall are located on the "shallowest" level (109). The entry into the plant is also hidden. The facility is only accessible via a lift shaft located in the corner of a ground floor barn which is concealed inside a "massive steel drum", ironically embossed with "a rusted and faded image of a cow and a sheep" (109). The idea that venturing inside these hidden facilities would "open our eyes" to the realities of meat production is made disconcertingly obvious in the description of the lift shaft opening: the concealed seam parts "like a vertical eyelid" (109) before enveloping its passengers and transporting them underground.

The slaughtering facility is also located underground at Ablach Farm, "three storeys below the ground" (210) on the Transit Level of the facility; there is nothing lower than this except the vodsel pens themselves. This subterranean location not only mimics the physical distance between industrial slaughter and the Western consumer, but it also alludes to how the public largely remain 'in the dark' about the violence necessary for them to continue consuming animal flesh. The progression into the depths of the earth, or "all those arm's-lengths under the ground" (210) as Faber poignantly describes, also connotes burial—both of the realities of slaughter and of literal bodies, and alludes to the violent or 'hellish' aspects of the abattoir, reminding the reader that the consumption of meat cannot occur "without the death of an animal" (Adams 2010, 66). These 'hellish' aspects are again alluded to in the dream Isserley has whilst driving towards the ruins of a medieval abbey where she often rests while out on the road seeking vodsel hitchhikers for her kind. She begins dreaming of the "ocean of sky" above the ruins before slipping down into a "deeper level", "through a treacherous crust of pulverulent earth" before landing in the "subterranean hell of the Estates" (Faber 2000, 117) (her former home). Here, she visualizes the dark centre of a factory, and is repulsed by a "giant concrete crater" filled with rotting plant matter, where "baggy diver's suits enslimed in black muck" were attached to hundreds of tubes and reeled in by "indistinct" mechanical agents (118). "'This,'" a guide explains, "'is where we make oxygen for those above'" (118); at this revelation, she screams herself awake. The subversion of the oxygen-making process and the bleak portrayal of its production is a powerful way of critiquing the process of industrialized slaughter. Isserley's imagining of the dark and grim realities and the hidden

---

'alien' operation can continue unhindered. However, the underground location also alludes to the idea of the realities of factory farming (and slaughter) not being easily seen or accessed by the general public.

'dirt' and death required in order to make oxygen for the public is a subconscious response to the same 'dirty' deeds involved in meat production; deeds which she knows about, but which she refrains from viewing or criticizing.

Despite her successful avoidance of the vodsel pens for some time, Isserley eventually finds herself in the depths of the plant after feebly accepting Amlis Vess' request to join him. It is here, during Isserley's first encounter of the vodsels in their confined environment,[6] where Faber's critique of factory farms is at its most powerful. Upon reaching the bottom level of the vodsel processing plant, Amlis and Isserley are immediately greeted by almost complete darkness, "as if they had been dropped into a narrow fissure between two strata of compacted rock with only a child's faltering flashlight to guide them" (166); the stench of "fermenting urine and faeces" also permeates the space (166). Other than a few "feeble infra-red bulbs", all Isserley can see are the "firefly glints of a swarm of eyes" swaying everywhere in front of them (166). Amlis then poignantly asks: "Do you know where the light is?" (166). Here, Faber encapsulates the contemporary relationship between the human subject, the animal object, and the factory farm in one interrogative passage. The darkness symbolizes the distance between the three and the 'gap' in consumer knowledge about factory farming practices, whilst the visibility of the eyes allude to the fact that despite this gap, the consumer is well aware of the animals' existence. Finally, Amlis' question symbolizes the apparent ease with which this distance and gap might be remedied—the light or 'information' is not hard to find. Thus, "do you know where the light is?" is a question directed towards consumers, and asks of them not only whether they are aware of what goes on inside factory farms, but whether or not they are intentionally ignoring those realities.

The question comes at a timely point in the novel, for what follows is a description eerily reminiscent of factory farm conditions. As Isserley turns on the light, a "flood of harsh light" fills the room and she begins to feel claustrophobic due to the tight and cramped conditions created to maximize the number of vodsels (168). During the excavation of this, the deepest level, she notes:

> ... the men had burrowed out no more of the solid Triassic rock than they absolutely had to [...]. The vodsel enclosures, a corona of linked pens all along the walls, took up almost the entire floor space; there was just enough room left down the middle for a walkway. (168)

---

6 She had only seen the pens once before, after they were newly constructed, and before any vodsels had been captured (Faber 2000, 168).

Isserley is then confronted by the grimy and cramped cages, "the wire mesh soiled, masked in places with the dark putty of faeces and other unidentifiable matter" (168) and the "stench and looming density of flesh, the humid ambience of recycled breath" (168–169). The doctoring of animal feed that occurs within CAFOs is also referred to; the monthlings are described as glistening with the "dark diarrhoea of ripeness [...]. Nothing which might cause the slightest harm to human digestion survived in their massive guts; every foreign microbe had been purged and replaced with only the best and most well-trusted bacteria" (169). Isserley effectively acts as a proxy for the Western consumer at this point in the narrative, and these descriptions begin to 'fill the gaps' in consumer knowledge concerning intensive farming operations. And whilst, at this point in the narrative, Isserley does not share Amlis' concern for the vodsels' health and welfare and is more preoccupied with "how hard she must constantly be working" (169), the descriptions of the vodsels' close confinement is still intensely provocative and disconcerting for the reader in that they must picture, not chickens, hogs, or cattle in these dire conditions, but fellow human beings, who, like those animal species, have been physically mutilated in order to produce more docile and profitable bodies.

The power of intensive confinement to further objectify animals is also made evident in Faber's description of the vodsel pens. As Isserley walks around the enclosures, she sees that:

> ...monthlings were huddled together in a mound of fast-panting flesh, the divisions between one muscle-bound body and the next difficult to distinguish, the limbs confused. Hands and feet spasmed at random, as if a co-ordinated response was struggling vainly to emerge from a befuddled collective organism. Their fat little heads were identical, swaying in a cluster like polyps of an anemone, blinking stupidly in the sudden light. (169)

This potent description demonstrates how the close confinement of animals causes them to be conceived as one immense meat-producing apparatus and that this conception eliminates any understanding of the animals as individual subjects. The lack of space and large number of animals causes them to seem "identical" and the removal of subjectivity renders them, collectively, "befuddled", "confused" and "stupid" (169), and thus unworthy of consideration by those profiteering from their objectification. Not only are animals viewed as objects that *can* be intensively reared in close confinement and away from their 'natural' surroundings, but the conditions inside the factory farm *add* to their objectification, by creating the illusion of a collective, meat-producing, Cartesian 'machine'.

Speak No Evil: Language as Camouflage

In the same way that the *physical* separation between the human consumer and the objectified animal aids in the perpetuation of Western carnism, the *linguistic* separation between the two parties also serves to uphold the current meat-centric status quo. The categories assigned to nonhuman animal species and the language used to describe the industrial processes they must endure, are part of a larger framework of discourses that support the species hierarchy and promote the consumption of certain types of animal flesh. These discourses contain—and promulgate—particular types of 'knowledge', 'facts', or 'truths', which are then used to create demand for meat products through pervasive marketing; to justify developments in meat production such as genetic modification and factory farming; and to euphemize the violence of slaughter and the disturbing realities and consequences of factory farm conditions and biotechnologies. Foucault's assertion that discourses are "practices that systematically form the *objects* of which they speak" ("The Archaeology of Knowledge" 32 emphasis added) is particularly pertinent in an analysis of discourses at work in the realm of Western meat production and consumption; for this precisely describes how discourses such as these transform sentient animals into commodified objects by those in power—those that quite often have a commercial interest in the commodities produced. Due to the authoritative quality of these various discourses, the hierarchies and ideas that are embedded within them become part of the dominant ideology; in this way, discourses and the 'knowledge' or 'truths' that they put forward, become naturalized, normalized, and consequently, 'invisible'. In addition, industrial "double-speak" with its "abuse of euphemism, nominalization, abstraction, presupposition, jargon, titles, and metaphor" (Coe 1998, 192–195) works to camouflage the disturbing aspects of meat production whilst also rendering the animal object, rather than subject. Faber's text illuminates the ways in which factory farm and slaughterhouse discourses and "double-speak" are used to keep certain animal species very much 'caged' in their place within the hierarchy of species and trapped within the confines of factory farms, slaughterhouses, and Western menus.

The objectifying terms assigned to animals in the meat production industry are reflected in Faber's narrative via the labels given to the vodsel males Isserley captures. Hunted vodsels, for example, are referred to as "specimens" (1); "transitionals" (169) are named so because they have not long been castrated and had their tongues removed; and when they have reached "monthling" status (169), the vodsels are considered ready for "processing" (216). These labels mimic the way animals are defined and objectified by terms such as 'stock', 'units', 'breeders', 'growers' and so forth inside the factory farm. In addition, the power of language to mask the realities of factory farm operations is revealed through Isserley's euphemistic descriptions of the acts carried out at Ablach Farm. For example, the docking of the vodsels' tongues that occurs on their arrival is twice described by Isserley as the vodsels being "seen to", similar to the way in which neutered animals are described as having been 'fixed'. This violent act is further down-played when she explains that the procedure "nipped any problems in the bud" (174); the appropriation of a colloquial phrase here (which carries with it a reference to non-sentient life—the 'bud' of a plant) hides the traumatic and painful corporeal transformation that the vodsels must undergo.[7]

The names of the areas within Ablach Farm also obscure the occurrences within them. Terms such as the "Transit Level" and the "Processing Hall" hide the realities of what actually occurs in those spaces in much the same way that phrases such as 'meat packing', 'meat works', or 'freezing works' do.[8] "Transit" refers to both the arrival of the vodsels and the removal of their genitalia and tongues, and their preparation for shipping once they have been "processed"; the latter, of course, refers to the site of their slaughter and butchering. In addition, the apparatus in which they are both sterilized and later killed, is called the "Cradle" (211); constructed from various pieces of farm equipment, it is as far from associations of birth, peaceful slumber, and nurturing as can be. These euphemisms and camouflaging labels are a direct reflection of the "deliberate and self-conscious" language used by the Western animal farming industry to both remove the unsavory nature of certain farming practices, and naturalize their existence (Luke 2007, 174). Thus, through these terms and descriptions, Faber draws our attention to the camouflaging properties of the language of the CAFO, such as 'grower sheds' and 'barns' for the buildings containing thousands of animals in close confinement, 'sow stalls' or 'gestation crates' for pens so small the sows within them cannot turn around, and the dissonance within the term 'factory *farm*' itself.

---

[7] Although this common conceptualization of plant life has also recently been questioned. See Michael Marder's *Plant Thinking: A Philosophy of Vegetal Life*, Columbia University Press (2013).

[8] 'Freezing works' or 'the freezer' are common colloquial terms for industrial abattoir and meat-packing facilities in New Zealand.

Furthermore, the way that the killing floor and slaughter itself is described by certain characters also references the way consumers currently talk about, or rather 'around', the practice of animal slaughter. For example, Isserley continues to adopt euphemistic rhetoric when she asks to witness the slaughtering process despite having acquired a kind of "blood-lust" after being sexually assaulted by one of the hitchhikers: "... are there any... any *monthlings* still to be... processed?", she asks co-worker Unser (original emphasis); "I'd love to see [...] to see the way you do it. The end product..." (216). Another passage depicts a vodsel being brought in to be 'processed': "The men merely nudged him with their flanks whenever he seemed about to stumble or deviate. They accompanied him: that was the word. They *accompanied* him to the Cradle" (217, emphasis added). These excerpts, associated with Isserley's experience of witnessing the killing of a vodsel, draw attention to the calculated adoption of euphemistic language with regards to the slaughterhouse. To emphasize his critique, Faber also contrasts the calculated euphemisms employed by the characters with potent and visceral third-person descriptions of slaughter. For example, Isserley watches Unser as he "slashed open the arteries in the vodsel's neck, then stood back as a jet of blood gushed out, steaming hot and startlingly red" (219).

   Faber also draws our attention to the power of language to create distance between the human consumer and the animal 'product' through his use of culinary, food and meat-related imagery. Throughout the text, things that we would usually consider to be 'inedible' are compared to various foods; the result of which, is often troublesome and disconcerting. In doing this, Faber shows us that language objectifies animals in the way that we consider them—and describe them—as "always-already meat" (Vint 2010, 24); as "units" of "stock" in a factory farm, or slaughterhouse killing line. He also points out that what we deem to be 'edible' is as much constructed by language and discourse as the human-animal divide. The most obvious example of this critique comes by way of the descriptions of Isserley's targets; humans ('vodsels') are made edible through language. Isserley wants to "size them up" (1), avoids "puny, scrawny specimens" (1) and looks instead for "fleshy" (3) males in their "prime condition" (4) that "make the grade" (4). One hitchhiker has "beefy hands" (20), another has a "meaty face" (119), and human clothes are "like layers of cabbage or radish" (212). Female 'vodsels' are also described with this type of imagery. Whilst watching television, Isserley notes the female vodsels hanging clothes out to dry, and how they "teetered on tiptoe, jumping like infants, their pink breasts quivering like jelly" (153). Isserley herself is also described in a similar fashion. In a shocking depiction of a sexual attack against her which features in the latter part of the novel (and leads to Isserley's desire to watch a vodsel being killed and processed), her assailant "began to knead her breasts with the hand that wasn't holding the knife, each breast in turn, repeatedly trapping the nipples between thumb and forefinger, rolling them like pellets of dough" (184). The critique in this passage is two-fold: not only does it illustrate the objectifying power of language, it also destabilizes the human-animal boundary by highlighting the entanglement of human and animal welfare—the disconcerting similarities between the language used to describe 'meat' and women is alluded to here.

Food-related imagery is also used elsewhere in the text. When Isserley and Eswiss (the farm's 'manager', whose body has—like Isserley's—been modified to appear more humanlike), drive around the property looking for vodsels who have escaped, they transverse "a massive pie-slice some three miles in parameter" (97). The chute that holds the vodsels before the removal of their tongues is described as "gleaming and elegant like a giant gravy boat" (211). Even language itself is presented as something edible, although aptly unpalatable; Isserley asks one hitchhiker "Where are you heading?" and the question "hung in the air, cooled like uneaten food, and finally congealed" (177).

This critique is enhanced when Faber presents us with the opposite: things that we usually deem edible are made unappealing, unpalatable, and even dangerous. When human ('vodsel') food is referred to in the text—the edible, becomes inedible, and disconcertingly so. For example, when Isserley stops at a petrol station she notices she has spilled chocolates in her car and that they "had found hiding places in every cranny like rotund beetles" (154). She then picks up the box and reads the ingredients:

> 'Sugar', 'milk powder', and 'vegetable fats' sounded safe enough, but 'cocoa mass', 'emulsifier', and 'lecithin' and 'artificial flavours' had a chancy ring to them. In fact, 'cocoa mass' sounded positively lethal. Her gut-reflex queasiness was probably Nature's way of telling her to stick to the foods that she knew. (155)

An every-day (and much loved) food such as chocolate is "defamiliarized"[9] when described in this manner and takes on a sinister or suspicious undertone—again, the power of language to make something palatable or not is illustrated. It also prompts the reader to consider how often they actually read and comprehend the ingredients listed on the food they consume, and

---

9 I refer here to Victor Shklovsky's use of the term—the reader is forced to consider the text "with an exceptionally high level of awareness" when familiar words or images are presented in an unfamiliar way (see Lemon and Reis *Russian Formalist Criticism* 1965, 5).

whether or not they know where that food comes from. Perhaps they are too often beguiled by the "labels decorated with cartoon pigs or cows..." (Otter 2008, 106).

Serving Up Subversion: *Under the Skin* and Beyond

Under the Skin is a prime example of the power of the fictional narrative to elucidate real-world ideologies, and is one of a number of contemporary novels to address and critique many of the contentious issues that exist within the realm of Western meat production and consumption.[10] These fictional 'human-animal' narratives are an effective critical avenue; due to the binary way in which we conceptualize 'human' and 'animal', fictional depictions of the latter inevitably confront and interrogate the notion of the former. "The animal", writes Baker in *Picturing the Beast*, is "frequently conceived as the archetypal cultural 'other'" and plays a "potent and vital role in the symbolic construction of human identity" in various contemporary instances (2001, ix); in other words, "when humans imagine animals, we necessarily reimagine ourselves" (Daston and Mitman 2005, 6). Fictional works provide readers and audiences with critical distance, or a means by which they may look "askance" in order to make the invisible, visible (Foer 2009, 29), thereby providing an avenue for reflection on aspects of daily life that have become entrenched and, consequently, remain unseen and rarely challenged. The continuing prominence of meat in Western diets and the discourses harnessed to reinforce the status quo, our relationship to the animals from which this meat derives, the issues surrounding its production (on the paddock, in the laboratory, and behind closed doors in the factory farm and slaughterhouse), the effect of meat consumption on interpersonal relationships, and the metaphorical and symbolic value of meat might be missing from mainstream advertisements, prime-time news bulletins and reality television, but they are not exempt from the arena of fictional narrative and critical distance.

---

10 Other recent publications which deal with some of the themes discussed in this thesis are Margaret Atwood's "MaddAddam Trilogy"—*Oryx and Crake* (2003), *The Year of the Flood* (2009), and *MaddAddam* (2013), Don LePan's *Animals* (2009), David Agranoff's *The Vegan Revolution... with Zombies* (2010), Vicki Pardoe's young adult novel *Cooped Up: A Factory Farm Novel* (2014) and David Duchovney's *Holy Cow* (2015).

Faber's text and other contemporary fictions which challenge various aspects of Western meat-industry are also significant in that, by virtue of their cross-genre categorization, they are literary boundary transgressors themselves. These narratives may be assigned to a number of categories simultaneously—'dystopian', 'horror' 'satire', 'science fiction', 'fantasy', 'genre fiction', 'literary fiction' and so on, not to mention the numerous sub-genre categories that may also be applicable; such cross-category classification only enhances their transgressive critical agenda. The close analysis of fictions such as these also address and counteract the "severely disabling effect that regimes of taste can have on the social transformative function of literature" (Armstrong 2008, 225); that is, their cross-genre transgressivity and critical agenda destabilize notions about what types of literature are 'worthy' of analysis and criticism. Issues of taste and *taste*, can thus be addressed simultaneously, providing enough critical fodder to satisfy the appetite of the most discerning literary and/or Human-Animal Studies critic.

References

Adams, Carol. *The Sexual Politics of Meat*. New York: Continuum, 2010.
Agranoff, David. *The Vegan Revolution With Zombies*. Portland, OR: Deadlite Press, 2010.
Armstrong, Philip. *What Animals Mean in the Fiction of Modernity*. London; New York: Routledge, 2008.
Atwood, Margaret. *Oryx and Crake*. London: Bloomsbury, 2003.
Atwood, Margaret. *The Year of the Flood*. New York: Random House Large Print, 2009.
Atwood, Margaret. *MaddAddam*. London and others: Bloomsbury, 2013.
Baker, Steve. *Picturing the Beast: Animals, Identity, and Representation*, Urbana, IL: University of Illinois Press, 2001.
Coe, Richard M. "Public Doublespeak, Critical Reading, and Verbal Action." *Journal of Adolescent and Adult Literacy* 42 (1998): 192–195.
Compassion in World Farming. http://www.ciwf.org.uk/.
Daston, Lorraine, and Gregg Mitman. eds. *Thinking With Animals: New Perspectives on Anthropomorphism*. New York: Columbia University Press, 2005.
Duchovny, David. *Holy Cow*. London: Headline, 2015.
Elder, Glen, Wolch, Jennifer, and Emel, Jody. "Race, Place, and the Bounds of Humanity". *Society and Animals* 6.2 1998): 183–202.
Faber, Michel. *Under the Skin*. Edinburgh: Canongate Books, 2000.
Fiddes, Nick. *Meat: A Natural Symbol*. London; New York: Routledge, 1991.
Foer, Jonathan Safran. *Eating Animals*. New York: Little, Brown and Co, 2009.

Glenn, Cathy B. "Constructing Consumables and Consent: A Critical Analysis of Factory Farm Industry Discourse." *Journal of Communication Inquiry* 28 (2004): 63–81.

Imhoff, Daniel. ed. *The CAFO Reader: The Tragedy of Industrial Animal Factories*. Healdsburg, CA: Watershed Media, 2010.

Johnson, Andrew. *Factory Farming*. Oxford; Cambridge, MA: Blackwell, 1991.

Lemon L., and Reis, M. *Russian Formalist Criticism: Four Essays*. Lincoln: University of Nebraska Press, 1965.

Le Pan, Don. *Animals*. Montréal: Véhicule Press, 2009.

Luke, Brian. *Brutal: Manhood and the Exploitation of Animals*. Urbana, IL: University of Illinois Press, 2007.

Otter, Chris. "Civilizing Slaughter: The Development of the British Public Abattoir, 1850–1910." In *Meat, Modernity, and the Rise of the Slaughterhouse*, edited by Paula Young-Lee. Durham, NH: University of New Hampshire Press, 2008.

Pachirat, Timothy. *Every Twelve Seconds: Industrialized Slaughter and the Politics of Sight*. New Haven: Yale University Press, 2011.

Pardoe, Vicki. *Cooped Up: A Factory Farm Novel*. Vicki Pardoe, 2014.

Potts, Annie. *Chicken*. London: Reaktion Books, 2012.

Scully, Matthew. "Fear Factories: The Case for Compassionate Conservatism—For Animals." In *The CAFO Reader: The Tragedy of Industrial Animal Factories*. Healdsburg, edited by D. Imhoff. CA: Watershed Media, 2010.

Vint, Sheryl. *Animal Alterity: Science Fiction and the Question of the Animal*. Liverpool: Liverpool University Press, 2010. Print.

CHAPTER 9

Down on the Farm: Why do Artists Avoid 'Farm'[1] Animals as Subject Matter?

Yvette Watt

Introduction

In June, 2000, an article appeared in the *New York Times* titled "Animals Have Taken Over Art, And Art Wonders Why: Metaphors Run Wild, but Sometimes a Cow Is Just a Cow". The article detailed a number of exhibitions that included major works with animals as subject matter with the author stating that "These days it seems every contemporary art exhibition must have its animal, dead or alive"(Boxer 2000). This trend has seen a steadily increasing number of artists taking an interest in animals and human-animal relations over the last couple of decades, in line with a significant reassessment of the status of animals and the rapid growth of Human-Animal Studies as subject matter for scholarly investigation across a broad range of disciplines.

This growth of interest in the animal subject within the visual arts is very welcome given that, in the shifting hierarchy of what is considered fashionable or worthy subject matter for artists, animals have for several centuries been relegated to a rather lowly position. However—and despite the reference to cows in the title of the article—relatively few of these artists use their work to examine the relationship between humans and the animals we farm and eat.

The issue of the under-representation of farm animals in animal themed/based artworks and exhibitions was something that first came to my attention in the early 2000s. At the time my own artwork was in a state of transition, slowly but surely becoming more overt in its commentary on the problematic relationship between humans and other animals. At the time I was completing a studio-based Master of Fine Arts[2] which proposed that farm animals, whose

---

1 I use the term 'farm' animal here to refer to those animals most commonly farmed for food, such as cows (both for meat and milk), chickens (for eggs and meat), pigs and sheep. I place the word 'farm' in inverted commas to indicate the problematic nature of defining these animals thus.
2 Watt, "Food for Thought" 2003.

form is continuously changed by human intervention, bridge the nature-culture dichotomy, as they are seen simultaneously as 'natural' and 'human-made.' The body of work I produced thus reflected upon the uncertain status of farm animals in contemporary Western society whereby they are perceived as both nature and culture, subject and object, populous but marginalized. It seemed that farm animals' low status as subject matter for artists was likely to be related to confusion over how to categorize them. However, there was not the scope in the written component of my MFA to interrogate this matter in any depth, and the issue of both *how* and *how often* farm animals are used/represented by artists has remained a concern as I have watched the increasing presence of animals in the visual arts. This chapter aims to address these concerns by demonstrating that more than a decade after I completed my MFA farm animals remain under-represented in the visual arts, and proposes reasons for this situation.

Artists and Animals: A Survey

In order to provide evidence that farm animals are under-represented in animal based artworks, and also to offer some real insight into possible reasons for the lack of farm animals as subject matter for those artists whose work is concerned with animals and human-animal relations, a survey was devised. Aimed at professional, contemporary visual artists who use animals in their artwork, the survey was distributed widely in June/July 2014 via email to selected artists, as well as being posted and shared on Facebook, and was also listed on H-Animal Net and in the Australian Animal Studies Group e-bulletin and email Forum. Judging from the large number of responses received, many people forwarded the survey on to other artists, as I had requested.

The survey consisted of 29 closed ended multiple choice questions, 5 open ended questions, one rank order question, and 3 five-point Likert scale type questions. Respondents were also invited at several points during the survey to provide additional comments and explanations. The survey was broken into five sections, with the first section devoted to questions seeking basic demographic information, as well as asking whether the respondents had companion animals and whether they had ever lived on a farm. The second section, "Animals and Your Art Practice," was the lengthiest, asking questions related to the nature of the respondent's art practice, as well as about the types of animals featured, how the animals were featured (including whether dead animals or body parts were used), how often animals were featured and reasons for featuring animals. This section also asked respondents to rate a list of animals on a sliding scale from 'natural' to 'not natural.' The section "Art and Animal Activism" included questions about the role of art in drawing attention to animals rights or animal welfare issues, while the section "Animals as Food," asked questions about the respondents' dietary choices. The last section, "Relations Between Humans and (non-human) Animals," asked questions related to the rights of artists, scientists, farmers and the general public to use animals, as well as asking respondents about their level of involvement in the animal protection movement.

Approximately 390 artists participated in the survey.[3] A general picture of these respondents can be gained from the following overview.

The survey respondents ranged in age from 18–84, with 80% aged between 25 and 59 years of age. 93% have a university degree or diploma. They are current residents of 25 different countries, although in terms of place of birth, they hail from 38 different countries. The majority live in the USA or Canada (37.5%), Australia or New Zealand (35%), or the UK (16%). 40% of the respondents grew up in large urban centers, 37% grew up in small urban centers, towns or villages, and 21% grew up in a rural/pastoral areas. They now reside mostly in large urban centers (57%) or small urban centers, towns or villages (32%) with only 16% of respondents currently living in rural/pastoral areas.

In terms of contact with animals, a surprisingly large 41% of the artists surveyed have at one time lived on a farm (although as the types of farms were not specified it is not possible to say how many of these respondents had the opportunity for regular interactions with animals raised for food production). 73.5% of the artists surveyed currently have a companion animal, while only one respondent has never had a companion animal.

Almost 80% of the respondents were women, a significant fact given that research shows that in Australia and the USA (for e.g.) the gender breakdown of visual arts practitioners is close to even, and is over-represented by men when it comes to the gendered distribution of works in museum and gallery collections and significant international exhibitions (The Australia Council for the Arts).[4] Interestingly, the gender imbalance in the survey respondents

---

3   A total of 470 responses were received. After responses from undergraduate students were removed (as they are not deemed to be professional artists prior to the completion of their degree) 412 responses remained, of which 23 answered only a small number of questions.

4   It should also be noted that men are statistically much better represented in the animal protection/advocacy movement in roles of authority, which is mirrored in the over-representation of male artists in in museum and gallery collections and in significant international exhibitions. There is a important issue to be interrogated here but there is not the scope in this chapter to do so.

reflects a similar gender imbalance reported across the animal protection/advocacy movement.[5]

One important motivation one might expect to figure in the decision of an artist to feature animals in their art is concern for animal welfare. One way such a motivation can be expressed is through dietary choices and support for animal activism. Approximately 48% of the artists said that they maintain some dietary restrictions, including 11% who are vegan and 18% who are vegetarian, which is a substantially higher percentage than for the general population.[6] These artists said the most important reason for this was their "Concern for the rights of all animals" (44%). A further 35% nominated "Concern for the welfare of intensively farmed animals," with 5% nominating the most important reason being "Concern for the environmental impact of farming animals". Eighteen people responded by nominating 'other' and writing a comment. The majority of these respondents stated animal welfare related concern/s, such as a preference for free range or 'ethically' sourced meat/animal products, were the main reason for dietary restrictions.

In regard to the respondents' political engagement, 32% consider themselves to be animal rights/welfare activists, while 50% do not. 22% stated that they are members of animal rights/welfare organizations, but only 13% regularly participate in animal rights/welfare related events.

How and Why the Artists Use Animals in their Work

It is pleasing to see that a substantial 75% of the artists who responded to the survey agreed that "Artists have an important role to play in the positive representation of animals." A further 61% agreed that "Art should be used to further the cause of animal rights" and 48% said that they had done so within their own art practice. It is, however, not clear how these artists define the cause of animal rights or how they use their art to further such a cause. It is also heartening to find that only 10% of respondents had produced artworks that harmed or killed animals, either intentionally or unintentionally. In addition, the responses to the open-ended aspect of this question asking what type/s of animal/s were harmed or killed, indicate that approximately a quarter these artists were not in fact culpable for harming or killing animals for their

---

5   This has been demonstrated in many analyses of the animal protection/advocacy movement. See for example Gaardner, "The Gender Question." This is also a matter worthy of further investigation.
6   A 2012 Gallup Poll indicated only 5% of Americans considered themselves vegetarian and only 2% vegan. This is consistent with other available data (Gallup 2012).

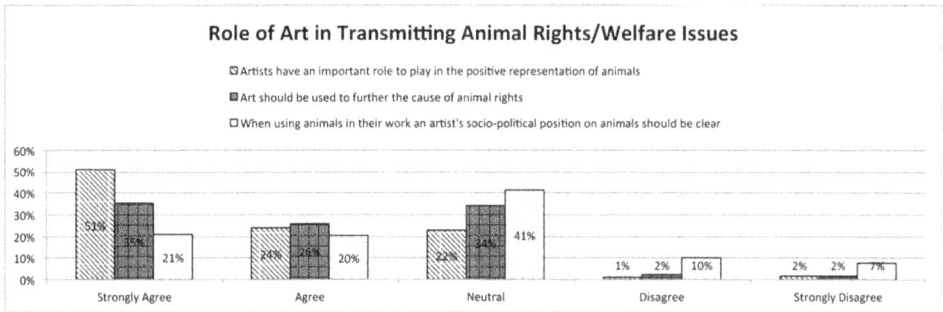

FIGURE 9.1 *The role of Art in Transmitting Animal Rights/Welfare Issues.*

Reasons for using animals	Strongly Agree	Agree	Neutral	Disagree	Strongly Disagree	Total responses
As symbols or metaphors standing in for someone or something else	23%	29%	20%	13%	15%	100%
For humour	5%	22%	25%	19%	29%	100%
To represent the natural world	40%	34%	15%	8%	3%	100%
As decoration	3%	12%	25%	22%	38%	100%
To celebrate the beauty of animals	44%	28%	17%	6%	6%	100%
To raise or draw attention to an environmental issue	46%	24%	19%	6%	5%	100%
To raise or draw attention to a social/political/ethical issue related to animals	49%	25%	16%	5%	6%	100%
To depict specific/individual animals (e.g. a portrait)	39%	28%	13%	11%	9%	100%

FIGURE 9.2 *"Please indicate the extent to which you agree with following reasons for featuring animals in your artwork."*

artwork. Rather, they stated such things as the animals being found dead (e.g. roadkill or donated taxidermy), or that the animals' deaths took place during documentary filming at a slaughterhouse, or even specified the animal based/sourced ingredients in conventional art materials and tools (e.g. hogs' hair brushes).

The survey also asked the artists to indicate to what extent they agreed with a range of reasons for including animals in their artwork. As can be seen from the table below "To represent the natural world" was the reason most strongly reported overall, with 74% agreeing and only 11% disagreeing. The least popular reason was "As decoration," followed by "For humour" which demonstrates that the majority of artists surveyed avoid trivializing animal subjects. However more than half of the artists use animals "As symbols and/or metaphors to stand in for someone or something else".

Artists and 'Farm' Animals

One of the most important questions in the survey in respect of this chapter was: "During the past ten years, what types of animals are most often featured in your artwork?"[7] The results show that about 40% of the artists who responded have used farm animals in their work, and of that number only 26% said that they used them "most often." On the other hand, almost 60% of the artists who responded have used wild living or native species, and 48% use these types of animal "most often"—and this figure does not include reptiles and amphibians, fish, or 'feral'/introduced animals. On one hand it is pleasing to see that farm animals feature as often as they do. However given that the animals used in agriculture are so numerous, so culturally entwined with human lives, and that human interaction with these animals, either alive or dead, is so ubiquitous, I would argue that they are radically under represented. For example a recent study has estimated that in terms of vertebrates, the biomass of domestic livestock now vastly exceeds that of wild animals (Smil 2011, 616–19). Further, as the surveyed artists were self-selecting and indicated a relatively high engagement with the ethico-political aspects of human-animal relations it is to be expected that these artists would be more likely to feature 'farm' animals than those artists who use animals in their work but who chose not to complete the survey.[8]

While it is relatively simple to demonstrate that farm animals are underrepresented by artists who are interested in animals and human-animal relations, it is a more complex matter to determine why this might be.

A key hypothesis for this research was that the absence of farm animals in contemporary art might reflect a broader societal perception that farm animals are not 'natural' or not sufficiently natural to embody a clear sense of pure animality, and that this confused status may be complicit in artists' general disinterest in these animals as subject matter. To test this supposition, the survey

---

7 Respondents could choose all that applied from the list provided, which was: companion animals; insects/invertebrates; reptiles and amphibians; farmed animals; working animals; aquaculture species; wild fish/shellfish; aquarium fish/shellfish; laboratory animals; wild living or native species; pest/feral/introduced animals; animals in entertainment industries; other.

8 For example, Karen Knorr responded via email to say she did not like surveys and was too busy, while Damien Hirst's assistant also replied via email to say he was too busy to respond.

Over the past 10 years, what type of animals are most often featured in your artwork?	Animals Featured	Animals Most Often Featured	Total
Farmed animals	153	102	389
Wild living or native species	233	185	389
Farmed animals	39%	26%	100%
Wild living or native species	60%	48%	100%

FIGURE 9.3 *"During the past ten years, what type/s of animals are most often featured in your artwork?"*

asked the artists to rate a variety of animals[9] on a scale of 'natural' through to 'not natural'[10] where 1 was 'natural' and 10 was 'not natural.' The results are significant and demonstrate clear differences in the perceptions of 'naturalness' for animals of different types/species. From the list provided, the animals perceived as 'most natural' were whales with an average rating of 1.49. Sharks and field mice both had an average rating of 1.52, while the animals perceived as least natural were Chihuahuas (4.68) and laboratory mice or rats (5.02). What is particularly instructive here are the ratings for farm animals. Chickens, dairy cattle, domestic pigs and beef cattle rated at the end of the 'not natural' scale with respective scores of 4.05, 4.35, 4.40 and 4.48. However, their wild relatives, jungle fowl, wild boar and water buffalo rated at the 'natural' end of the scale at 1.77, 1.83, 2.09 respectively—immediately after whales, sharks and field mice.

The data from this question is thus telling. As we have seen, the artists surveyed demonstrated a clear preference for using non-domestic animals in their artwork, with the most strongly reported reason they use animals in their artwork being to represent the natural world, and further, that the animals most commonly used are wild-living and/or native species. Given that so many of the surveyed artists rated farm animals at the 'unnatural' end of the scale, it would seem that the perception of impure animality and/or the 'unnaturalness' of farm animals is closely related to their low representation as subject matter for artists. But the survey suggests additional reasons for this underrepresentation of farm animals.

---

9   The full list of animals was: German Shepherd Dog; Chihuahua; Domestic Cat; Feral or Stray Cat; Mouse or Rat in the Built Environment; Field Mouse; Laboratory Mouse or Rat; Water Buffalo; Dairy Cow; Beef Cattle; Wild Boar; Domestic Pig; Jungle Fowl; Chicken; Whale; Koi; Shark; Trout.
10 Where not natural was defined as 'culturally constructed'.

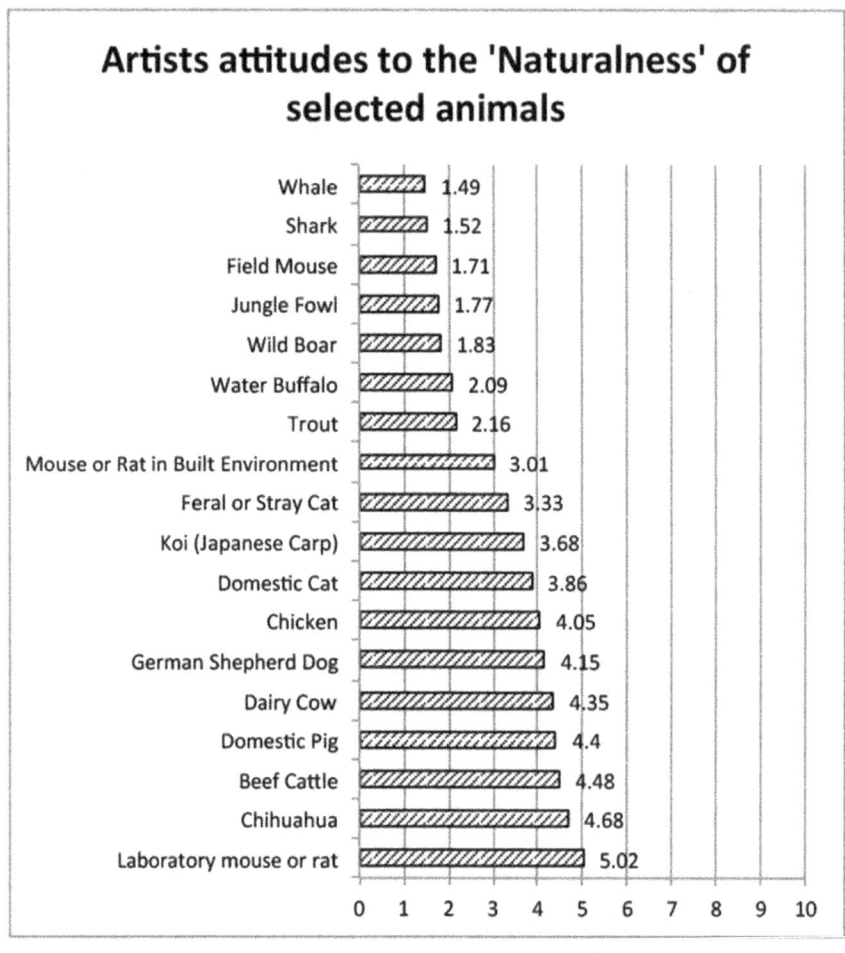

FIGURE 9.4 *"Please indicate where you would locate the following animals on a scale ranging from 'natural' to 'not natural' where 'not natural' means culturally constructed."*

As noted, while almost 30% of the artists who responded to the survey are vegan or vegetarian more than 70% continue to eat meat. While some state a preference for purchasing/eating meat from animals raised in more ethical ways, a significant proportion still choose to eat animals. Thus, presumably, a substantial majority of the artists essentially perceive farm animals to be commodities created by humans, for human use. This cannot help but affect their attitudes toward these animals and therefore affect their choices about using—or more to the point, not using farm animals as subject matter for artwork. This is reflected in the fact that the survey shows that of those artists who are vegan or vegetarian, 45% feature farm animals most often in their artwork.

However, of those who are not vegan or vegetarian only 23% feature farm animals most often. Thus the artists who are vegetarian or vegan are significantly more likely to use farm animals in their work, and twice as likely as the non-veg*ns to employ them as the main type of animal subject for their artwork.

Animals, Art and Ethics

What is interesting in the context of these statistics is the high percentage (75%) of the surveyed artists who believe art has an important role to play in the positive representations of animals, with 72% saying that they use their artwork to draw attention to social, political or ethical issues related to animals. Artist Bryndís Snæbjörnsdóttir, who works collaboratively with Mark Wilson, expands upon the significance of using animals in this context:

> For me and many other artists that engage in socially engaged art, art is a serious tool of investigation and a powerful lever to instigate social change. It is therefore impossible to read the question "Is it ethical to use animals in art?" without thinking "Is it ethical to use animals in science?," "Is it ethical to use animals in cooking?" (Baker 2013, 14)

In light of Snæbjörnsdóttir's comment it is instructive to see that of the artists who responded to the survey, 30% believe "Farmers should be able to use animals in any way they see fit so long as it causes the animals minimal harm or distress." They gave artists less permission in this regard, with only 25% believing artists should have the same right to the use of animals, but just 23% thought that scientists should have the same right and only 16% thought people in general should have this right. The fact that farmers are given more permission to use animals than artists, scientists or the general public provides further evidence that a substantial proportion of artists who use animals in their artwork see farm animals as commodities. More troubling, perhaps, the responses to this question suggest that these artists also believe they have more right to exploit animals than do the general public or even scientists. In his recent essay on "The Practice and Ethics of the Use of Animals in Contemporary Art" Joe Zammit-Lucia addressed this very issue, stating that "With freedoms come responsibilities. And there is no reason why artists should, in any way, be considered individuals afforded special privileges that are denied to others" (Zammit-Lucia, 7). It would seem that not all the surveyed artists would agree, a matter which is supported by artist and Animal Studies scholar Steve Baker who has stated that he is "perilously close to arguing that artists should be allowed certain freedoms that scientists should not be allowed!" (Baker and Gigliotti 2006, 37).

Farm Animals as Subjects and Objects

In discussing John Berger's examination of the empowered gaze in his 1980 essay "Why Look at Animals?" Steve Baker concludes that "only by understanding who has power over the image can we begin to elaborate a worthwhile cultural history of the animal" (Baker 15). Now more than ever in the history of humankind, our experience of the world is profoundly influenced by the overwhelming number of images that bombard our everyday lives and images of animals affect us both in their presence and in their absence. Our most common experience of the farm animal is on a dinner plate, but this is an absence, as much as it is a presence. Carol J. Adams pointed this out so succinctly when she spoke of how:

> Behind every meal of meat is an absence: the death of the animal whose place the meat takes. The "absent referent" is that which separates the meat eater from the animal and the animal from the end product. The function of the absent referent is to keep our "meat" separated from any idea that she or he was once an animal, to keep the "moo" or "cluck" or "baa" away from the meat, to keep the some*thing* from being seen as having been some*one*. (Adams 2000, 14)

The notion of animals as the "absent referent" extends to the art world, whereby the animals may seem to be present, but too often are there to represent something, or someone else, as is evident by the fact that more than half the surveyed artists use animals "as symbols or metaphors standing in for someone or something else" (see figure 2). As Jonathan Burt has noted, this "standing in for" leads to a situation where "the animal figure is split by a tension between signification and designification, between looking *at the animal* and *through the animal* to something else, whether these are abstract concepts, metaphorical meanings or expressions of inner psychic states" (Burt 2011, 164).

The status of the farm animal as an 'absent referent' has been aided by the increasing industrialization of farming practices and the urbanization of Western societies over the past few centuries, which has resulted in a removal of farm animals from our day-to-day lives and a consequent separation in the relationship between production and consumption of animals for food (Franklin 1999, 126). As Adrian Franklin notes "... people of the twentieth century have become spatially detached from the animals they consume and emotionally reluctant to recognize the embodied nature of meat foods" (Ibid). The result is that the animal becomes an object, a commodity that has been created by humans for a specific purpose and, as such, very little consideration is given to the animal's self-interest. In a legal sense, animals are indeed objects that can be bought and sold, housed, fed and slaughtered without much concern. In his book from 2000, *Dearest Pet*, Midas Dekkers recounts an example of a recent court case from Saxony where a man was on trial, having raped a cow and a nine month old calf. He raped the calf first, but when he tried to rape the cow she kicked him, knocking him to the ground. Enraged, he took a manure fork and forced it into the anus of both the cow and the calf. The cow died shortly afterward and the calf had to be euthanized the following day. The accused man was sentenced to two years and three months for damage to property (Dekkers 2000, 126). Once understood primarily as property the animal's essential worth is diminished to that of a mere commodity—an object to be bought and sold and used as desired with little or no consideration given to the consequences for the individual animal. As Professor Lesley Rogers, a neuroscientist specializing in animal behavior (in particular, the development of behavior in chickens), points out; "The animals we eat or exploit in other ways tend to be devalued relative to [other animals]" (Safe 2002, 25).

The object status of animals is also strongly evident in the art world. As Joe Zammit-Lucia has observed, "contemporary art has entered a stage where the animal has been degraded to the status of mere artistic material." (Zammit-Lucia, 1 abstract). This reification is even more pronounced when it comes to farm animals. For example, in the same essay, Zammit-Lucia discusses a sculptural installation work of Chinese artist, Cai Guo-Chiang titled *Head On*, which is comprised of a pack of 99 wolves. Rather than being actual taxidermy wolves, the life-sized and lifelike animals are made of resin, hay, metal wire and sheepskin. While Zammit-Lucia comments on the irony of turning sheep into wolves (which as he notes is apparently not an intended reading of the artwork),[11] he does not address the fact that, despite the use of animal skins, this widely seen and admired artwork has been the subject of little or no controversy. Had the 99 wolves been made of the skins of actual wolves, rather than the skins of sheep, one cannot help but wonder if response to this work would have been so overwhelmingly positive.

Further evidence of the object status accorded to farm animals in art is to be found in an exhibition catalogue from 2002 where American artist, Mark Ryden writes about an Austrian artist, Flatz, who "... made the news when he

---

11   Ibid p. 3.

dropped a dead cow from a helicopter in Berlin. An animal loving teenager attempted to legally stop the performance [but] the court rejected the complaint as the cow had the legal status of food" (Ryden 2002).[12] In considering this issue Ryden asks "At what exact point does the animal cross the line and become meat?" (Ibid). This is an important question as it underpins the object status of farm animals that makes them suitable for artists to use as artistic material, but not necessarily as subject matter—at least not in a manner that recognizes and respects their sentience and individuality.

The use of dead animals and animal body parts as both subject matter and physical material for artists has a long history, and one that has continued into the 21st century.[13] As Burt points out "artists who work with the animal body do not all do so for ethical or proanimal reasons, so that even in contemporary art we cannot assume a direct address to the animal" (Burt 2011, 190). Indeed, I would add that when an artist uses an animal carcass and/or flesh as material it is rare that they do in fact do so for ethical or proanimal reasons, and so it is no coincidence that farm animals are often the animal of choice for such use given their status as meat as producing commodities. This allegation is supported by the survey results.

When the surveyed artists were asked for the reasons they feature animals in their artwork, a substantial 72% of artists said that they use their artwork "To raise or draw attention to social, political or ethical issues related to animals." 73% also nominated "To raise or draw attention to an environmental issue." However, as noted earlier, only 48% had used their work to further the cause of animal rights/welfare. There is a telling gap here. There is a clear statistical preference by the surveyed artists to use non-domestic animals, and to also use animals to represent the natural world. Given this and the strong emphasis on using their animal-based artwork to promote environmental concerns, it seems likely that the majority avoid using their work to engage with the exploitation of the animals they choose to eat. Without doubt, the animals which are most commonly the subject of animal rights campaigns are farm animals, due to the vast numbers of these animals raised and killed each year, and the extent of the suffering they endure as part of standard farming and slaughtering practices. As such the farm animal is a highly politicized beast. Given that a high percentage of the respondents eat meat, it is reasonable to assume that featuring farm animals in their artwork in ways that respects these animals' sentience and individuality would create a sense of personal conflict in these artists.

---

12   Section 32 (no page numbers listed).
13   This is discussed in some detail by Jonathan Burt in "Animals in Visual Art" 2011.

So, as I have established, the majority of the surveyed artists continue to eat meat, and in doing so are compelled to avoid taking a political stance in relation to what we might term 'broad animal rights,' (meaning the application of equal consideration to all animals regardless of species) in their artwork. They do this by mostly choosing not to feature farm animals in their work. This is in contradiction to academic and artistic responses to other social justice issues such as race, gender and sexuality, whereby concern about the exploitation/persecution of the parties of concern is all encompassing rather than limited to particular subsets. It is also at odds with a rethinking of human-animal relationships in other disciplines where there is an increasing emphasis on the importance of foregrounding ethical and political issues. As Kay Anderson points out:

> The human-animal divide is increasingly being problematized in the human sciences, along with other conceptual distinctions of mind-body/male-female that over time have interacted with it. Such dualistic thought is under challenge by postcolonial and feminist scholars [...] The study of animals has thus been brought into a culture/society framework from which it has long been excluded [...] (Anderson 1997, 466)

In essence, the politicization of social justice issues is generally accepted as inherent, and artists and academics consider their engagement with the politico-ethical consequences of exploitation/persecution in such fields unavoidable, if not essential. Given this, we must ask why is it that the majority of artists who feature animals in their work, apparently for pro-animal reasons, continue to eat meat and thus seriously limit their engagement with the ethics of human-animal relations? Canadian political philosopher Will Kymlicka addressed this matter in a recent public lecture (Kymlicka 2014). In his presentation Kymlicka quoted Canadian author Blaire French who referred in his book *The Ticking Tenure Clock* to animal rights advocates as "orphans of the left."[14] The proposition here is that supporters of other left-leaning causes such as those relating to the natural environment and various human rights issues, tend to support all of these causes. However, the same cannot be said for the cause of animals rights, with the majority of the supporters of these other causes of the 'left' drawing a line when it comes to supporting animal

---

14 In an email to the author of 17 September 2014 Kymlicka explained that: "The term 'Orphans of the Left' comes from Blaire French's academic novel, *The Ticking Tenure Clock* (SUNY Press, 1988). The fictional protagonist in the novel hopes to get tenure on the basis of her forthcoming book on animal rights activists entitled 'Orphans of the Left.'"

rights. This same pattern is evident with the artists surveyed for this essay, who appear to on one hand use their work to promote issues of concern for environmental destruction and species loss while continuing to eat meat and other animal products, even though we hear more and more regularly of the serious negative effects of animal agriculture on the environment.[15] In his lecture Kymlicka stated that virtually every society is still based on the proposition that animals are property and that humans have a right to use animals for our needs, despite the fact that many also believe animals deserve at least some protection.[16] This inherently flawed desire to, on one hand 'protect' animals from harm, while at the same time allowing exploitation of them for our own purposes is at its most profound when it comes to the breeding, raising and slaughtering of animals for food, whereby in most cases the entirety of their births, lives, and deaths, are under human control, with little or no regard given to the desires of the individual animal. Australian Human-Animal Studies scholar, Dinesh Wadiwel provides a rationale for the persistence of this apparently hypocritical position:

> The law becomes an expression of a perpetual form of victory, which guarantees a continuing freedom for the victors. Freedom, in this sense, is not connected to equality; on the contrary, it conveys the opposite sense—in Foucault's words 'freedom is the ability to deprive others of their freedom'. This in turn enshrines a form of law that guarantees a continual pleasure for the victors—a freedom of unending satisfaction... In other words, law guarantees an unending flow of pleasures... (Wadiwel 2009)

In referencing Wadiwel (2009, 287–8), Kymlicka suggests that the flawed logic that insists on simultaneously wanting to eat animal products while comforting ourselves that the animals we eat are not suffering (or at least not suffering too much) is driven by the "flow of pleasures" that come to humans through animal exploitation, and we are further exonerated from concern at the consequences of this "flow of pleasures" due to these pleasures being culturally sanctioned. It would appear that despite their interest in animals and human-animal relations, the attitudes toward animals, and in particular farm animals,

---

15   For example, Food and Agriculture Organization of the United Nations. "Livestock's Long Shadow," and Carus, "UN Urges Global Move to Meat and Dairy-free Diet" (2014).

16   According to a recent Gallup Poll 96% of Americans say that animals deserve at least some protection from harm and exploitation, while just 3% say animals don't need protection "since they are just animals." (Gallup 2003).

by the majority of the surveyed artists are colored by their embracing of this culturally sanctioned "flow of pleasures."

Conclusion: Where to from Here?

This essay has established a clear connection between the commodification and consequent trivialization of farm animals and the fact that they are both under-represented by artists who feature animals in their work, and over-represented when it comes to their bodies and body parts being used as mere artistic material—from hogs' hair brushes to the use of animal bodies and body parts in artworks such as those produced by Damien Hirst, Hermann Nitsch or Cai Guo-Chiang to name but a few. I have demonstrated that the perception of farm animals as less-than 'natural' is complicit in their devaluation as fitting subject matter for the majority of the artists who responded to the survey. As Stephen Eisenmann observed, back in the 18th century Buffon argued that "the domesticated animal is far less noble than the wild one..." (Eisenmann 2013, 153). Unfortunately this attitude persists, both in general society and in the visual arts, to the detriment of those animals we farm for food.

Nonetheless there are some very positive messages to be taken away from the survey. The large number of responses received tells us that there are a substantial—and it would seem growing—number of artists who feature animals in their work.[17] Their willingness to undertake the survey suggests that these artists treat the use of animals in art with great seriousness, which is further evident in the time many took to add comments where this option was available to them, and is supported by the substantial amount of correspondence the survey provoked. Further, for the vast majority of respondents, there is a very real concern for the welfare of animals and the preservation of species. They see art as an important tool for engaging with the social, ethical, environmental, and political issues surrounding animals and human-animal relations, and many of them address such issues either directly or indirectly in their work.

Less positively, more than half of the surveyed artists find it acceptable to use animals as symbols or metaphors, to stand in for someone or something else, and whereas almost three quarters of the artists agree with using animals

---

17   The survey also generated a large amount of email traffic from artists interested in knowing more, passing on details of other artists, and in the case of several high profile artists including Hubert Duprat, Karen Knorr and an assistant of Damien Hirst, letting me know that they were not willing/able to complete the survey.

in their work to represent the natural world, to celebrate the beauty of animals or to raise or draw attention to a social/political/ethical issue related to animals, only two-thirds agree to depicting specific/individual animals as a reason to use animals in their work. This is significant, as it results in 'looking through' the animal, thus denying his or her individuality, sentience and self-interest.

In addition, given that 32% of respondents identify as activists and 61% agreed that art should be used to further the cause of animal rights/welfare (and indeed, close to 50% said that they *had* used their art for this purpose), the survey results are likely to be skewed toward artists who engage strongly with the ethical considerations of human-animal relations. On one hand this may give a false reading of the broader use of animals in art, but on the other hand if this is the case, it is significant that even among such ethically/socially engaged artists farm animals still rate so lowly as subject matter.

As I have observed here and in a previous essay,[18] the apparent concern felt by artists for animals is often more to do with the likelihood of a human-caused ecological catastrophe, and an accompanying anthropocentric concern about what such a disaster would mean for *humans* as much as what it would mean for animals. As Josephine Donovan and Carol J. Adams state:

> We could respond that many efforts on behalf of animals will qualitatively improve humans' living conditions as well, which is likely to be the case. But such an argument reduces analysis of interspecies oppression to a human-centered perspective. Yes, in terms of reducing environmental degradation, challenging the mal-distribution of food because of the squandering of food resources in the production of 'meat', and preventing human diseases associated with eating animals, such as heart disease and certain forms of cancer, it is true that it is in humans' interest to be attentive to and challenge animal exploitation. But these responses concede an insidious anthropocentrism while trying to dislodge it. (Adams and Donovan 1995, 4)

In terms of the methods used in the survey, it is important that I acknowledge that I am basing this research on the representation of very broad categories of animals, rather than on species or even breeds. Nonetheless it is significant given how ubiquitous farm animals are, at least in the form of their flesh, milk, eggs and skins, that they are so under-represented in the visual arts, and further, that despite the high proportion of survey respondents who indicate that

---

18   Watt, "Making Animals Matter" (2011).

they engage with the ethical issues surrounding human-animal relations, so few use their artwork to tackle the relationship between humans and the animals we farm for food.

It is also important to acknowledge a growing call from Human-Animal Studies scholars and artists for the art world to engage with a more ethical consideration of the use of animals in art, including Joe Zammit-Lucia's recent and very useful essay "Practice and Ethics in the Use of Animals in Art", which I have referred to in this essay a number of times. Of further interest in this regard is the decision by the American College Art Association to conduct a survey to determine a set of guidelines for "The Use of Animal Subjects in Art" (College Art Association). However, as exemplified in the wording of the full title of these guidelines ("A Statement of Principles and Suggested Considerations") there remains a hesitance in being too forthright about what should and should not be allowable when it comes to the use of animals in art, based on a fear of restricting artistic expression. Unfortunately, as Zammit-Lucia observed "The ethics of the use of animals in art have received considerably less attention in the mainstream art world" (Zammit-Lucia, 6). He is right to observe that these issues must be taken more seriously and discussed with more enthusiasm by the mainstream art world. As with other social justice issues, artists have a key role to play not only in addressing human-animal relations, but also in taking ethical considerations into account when doing so. Ironically, despite his essay expressing a fairly strident concern for animal abuse at the hands of artists, Zammit-Lucia also appears to soften his stance when it comes to establishing boundaries for artists in terms of what is and isn't acceptable when using animals. In quoting a statement I made in an essay from 2011,[19] he writes:

> Of course for some the answers may be perfectly clear. For an animal rights activist living in a world that is black and white, "causing an animal to suffer or die in the name of art is always unjustifiable." On the other side of the discussion there will no doubt be those who will argue that, provided their behaviour falls within the law, there is no justification for curtailing artists' freedom of expression in any way and that any such attempts would be unconstitutional. These are irreconcilable positions that simply lead to polarization rather than further enlightenment. (Zammit-Lucia, 7)

---

19   Zammit-Lucia is here referring to my essay "Artists, Animals and Ethics," in *Antennae: The Journal of Nature in Visual Culture* 19 (Winter 2011): 66.

In the same issue of *Antennae* animal activist and writer, Ashley Fruno, raised the matter of animal abuse in the name of art. Discussing the work of Hermann Nitsch and Jesse Powers (who tortured and killed a stray cat in 2001 for a college art project, apparently to address issues of hypocrisy and greed), Fruno stated:

> The abuse and torture of an animal can never be justified. By engaging in the exploitation and torture of animals to produce "art," Nitsch and Powers are evidently supporting the so-called hypocrisy which they are attempting to combat. Though their aim is to also initiate a conversation regarding society's hypocritical attitude towards animals, the methods which are used invoke a reactive response from the community, which contradict that. This response directs focus on the artists' lack of compassion and morals, rather than the relationship between their actions and those which occur daily in society. (Peled 2011, 58–9)

Fruno and I have closely aligned positions on the use of animals by artists, and our perspectives are shared by a growing number of artists and academics. Nonetheless I contend that I certainly do not see the issues as simply black and white, as asserted by Zammit-Lucia, but I also refuse to allow that because something is deemed legal—i.e. is culturally sanctioned—it must therefore be accepted without question. The issue has also been debated at some length in a published discussion between Steve Baker and Carol Gigliotti. These two scholars (and artists) differ markedly in their positions, with Baker effectively arguing for artistic freedom unencumbered by any potential limitations that taking an ethical stance might require, while Gigliotti feels strongly that artists should accept an ethical responsibility for their work and refuses to be circumspect about her opinions on the matter, stating that "I have decided that time is too short, for me, and for the planet, not to speak directly about these issues" (Baker and Gigliotti 2006, 40). It is important to point out here that restrictions on artistic freedom come with a similar set of concerns as those regarding restrictions on freedom of speech, but in neither case does this allow for total freedom to do, or say, whatever one likes unimpeded by any legal, ethical, or moral boundaries. As for any restriction on artistic freedom implying by necessity a restriction on creativity, there is much to suggest the opposite to be true—boundaries are in fact essential to the creative process, and within their art practices artists contend daily with any number of boundaries, including financial, thematic and ethical, sometimes imposed externally, as well as self-imposed.

As addressed earlier, images are powerful, not only reflecting the social milieu in which they are made, but also aiding in affecting social change. Thus *how* artists use animals and *which* animals are used are serious matters with very real repercussions, both for those individual animals used by artists in ways that directly affect them, and for the broader implications for societal attitudes toward animals that artists are complicit in affecting. Historically, artists are great questioners of culture and it is imperative that more artists question the nature of contemporary human-animal relations, particularly when it comes to those billions of animals who are raised and slaughtered for food annually, and especially those artists who feature animals in their work. This essay thus closes with a call for those artists, curators, and critics who are interested in animals and human-animal relations to make the effort to question at depth their own ethical position when it comes to their relationship to animals, and also to question the social context that culturally sanctions animal abuse, especially for those animals deemed commodities. There is no better place to start this inquiry than by questioning the perception and treatment of those animals unfortunate enough to have been categorized as food, and by making art that is, at the very least, informed by such conscious ethical engagement, and that might even encourage the viewer to undertake a similar self-questioning.

References

Adams, Carol J., and Josephine. Donovan. eds. *Animals and Women*. Durham: Duke University Press, 1995.
Adams, Carol J. *The Sexual Politics of Meat*. New York, Continuum: 2000.
Anderson, Kay. "A Walk on the Wild Side: A Critical Geography of Domestication." *Progress in Human Geography* 21.4 (1997): 463–85.
Australia Council for the Arts. "Artfacts," Accessed September 5, 2014. http://artfacts.australiacouncil.gov.au/visual-arts/creation/.
Baker, Steve. *The Postmodern Animal*. London: Reaktion Books, 2000.
Baker, Steve and Carol Gigliotti. "We Have Always Been Transgenic." *AI & Society* 20 (2006): 35–48.
Baker, Steve. *Artist/Animal*. Minneapolis University of Minnesota, 2013.
Berger, John. "Why Look at Animals?" In *About Looking*. New York: Pantheon Books, (1980): 1–26.
Boxer, Sarah. "Animals Have Taken Over Art, and Art Wonders Why: Metaphors Run Wild, but Sometimes a Cow Is Just a Cow." *New York Times*, June 24, 2000. Accessed September 3, 2014. http://www.nytimes.com/2000/06/24/arts/animals-have-taken-over-art-art-wonders-why-metaphors-run-wild-but-sometimes-cow.html?pagewanted=all&src=pm.

Burt, Jonathan. "Animals in Visual Art from 1900 to the Present." In *A Cultural History of Animals in the Modern Age*, edited by Randy Malamud, 163–94. Oxford: Berg, 2011.

Carus, Felicity. "UN Urges Global Move to Meat and Dairy-free Diet." *The Guardian*, June 2, 2010. Accessed December 12, 2014. http://www.theguardian.com/environment/2010/jun/02/un-report-meat-free-diet?CMP=share_btn_fb.

College Art Association. "Standards and Guidelines. The Use of Animal Subjects in Art: Statement of Principles and Suggested Considerations. Accessed November 11 2014. http://www.collegeart.org/guidelines/useofanimals.

Dekkers, Midas. *Dearest Pet: On Bestiality*. London: Verso, 2000.

Food and Agriculture Organization of the United Nations. "Livestock's Long Shadow: Environmental Issues and Options." Rome, 2006. Accessed July 8, 2014. http://www.fao.org/docrep/010/a0701e/a0701e00.HTM.

Eisenmann, Stephen F. *The Cry of Nature: Art and the Making of Animal Rights*. London: Reaktion Books, 2013.

Franklin, Adrian. *Animals and Modern Cultures*. London: Sage, 1999.

Gaardner, Emily. "The 'Gender' Question of Animal Rights: Why are Women the Majority?" Accessed September 20, 2014. http://citation.allacademic.com//meta/p_mla_apa_research_citation/1/0/3/8/6/pages103868/p103868-2.php.

Gallup. "In U.S., 5% Consider Themselves Vegetarians." July 26, 2012. Accessed 5 September 5, 2014. http://www.gallup.com/poll/156215/consider-themselves-vegetarians.aspx.

Gallup. "Public Lukewarm on Animal Rights." May 21, 2003. Accessed September 20, 2014. http://www.gallup.com/poll/8461/public-lukewarm-animal-rights.aspx.

Kymlicka, Will. *Towards a Muliticultural Zoopolis: Animal Rights, Race and the Left*. Public Lecture, University of Tasmania, August 11, 2014.

National Museum of Women and the Arts. "Get the Facts." Accessed September 5, 2014. http://nmwa.org/advocate/get-facts.

Peled, Zoe. "Discussing Animal Rights and the Arts." *Antennae* 19 (Winter 2011): 53–61.

Ryden, Mark. *Bunnies and Bees*. (Exhibition catalogue). California: Grand Central Press, 2002.

Safe, Mike. "The Cluck Stops Here." *The Weekend Australian Magazine*. May 25–26, 2002.

Serpell, James. *In the Company of Animals*. Cambridge: Cambridge University Press, 1986.

Smil, Vaclav. "Harvesting the Biosphere: The Human Impact." *Population and Development Review* 37 (December 2011): 613–636

Wadiwel, Dinesh. "The War Against Animals: Domination, Law and Sovereignty," *Griffith Law Review* 18.2 (2009): 283–97.

Watt, Yvette. *Food for Thought: A Visual Investigation of the Nature-culture Dichotomy as Manifested in 'Farm' Animals*. Unpublished Masters of Fine Art thesis, 2003.

Watt, Yvette. "Making Animals Matter: Why the Art World Needs to Rethink the Representation of Animals." In *Considering Animals: Contemporary Studies in Human-Animal Relations*, edited by Elizabeth Leane, Carol Freeman and Yvette Watt, Ashgate, Surrey, 2011.

Watt, Yvette. "Artists, Animals and Ethics." *Antennae: The Journal of Nature in Visual Culture* 19 (Winter 2011): 66.

Zammit-Lucia, Joe. "Practice and Ethics of the Use of Animals in Contemporary Art." *Oxford Handbooks Online*. Oxford University Press. Accessed June 30, 2014. http://www.oxford handbooks.com.

CHAPTER 10

The Provocative Elitism of 'Personhood' for Nonhuman Creatures in Animal Advocacy Parlance and Polemics

Karen Davis

It is increasingly being recognized that other animals besides humans have complex mental lives. They not only can suffer pain, injury, and fear, but they are intelligent beings with rich and varied social and emotional lives, capable of decision-making, empathy and pleasure. Based on the wealth of evidence, the great apes in particular—gorillas, chimpanzees, and orangutans—have been singled out for showing a range of mental capacities demanding that the moral boundaries we draw between them and ourselves must be changed. In 1993, *The Great Ape Project*, edited by Paola Cavalieri and Peter Singer, argued that the "community of equals" should be extended to include "all great apes" (Cavalieri and Singer 1993, 4). Currently, the Nonhuman Rights Project, founded by attorney Steven Wise, is working through the courts to change the common law status of some nonhuman animals from mere 'things,' which lack the capacity for legal rights, to 'persons,' who possess the fundamental rights of bodily integrity, liberty and other legal rights to which "evolving standards of morality, scientific discovery, and human experience entitle them" (Wise 2014, 1). While focusing on legal rights for chimpanzees, the Nonhuman Rights Project suggests that expanding the moral and legal community to include these animals could initiate a larger break in the species barrier. For nonhuman animals, Wise says, "The passage from thing to person constitutes a legal transubstantiation."

While this is an exciting prospect, some animal advocates worry that the Great Ape Project and the Nonhuman Rights Project could reinforce the very attitudes and assumptions of elitism that have caused so much misery to animals in the world. In both projects, humans are at the top of the scale and the great apes follow. Below them some other mammals await consideration, and further down some species of birds may appear. Reptiles, fish and insects are either absent or at the bottom. In Peter Singer's book *Rethinking Life and Death*, the only beings who qualify conclusively as 'persons' are the great apes, although he say that whales, dolphins, elephants, monkeys, dogs, pigs, and other animals "may eventually also be shown to be aware of their own existence over time and capable of reasoning. Then they too will have to be considered as persons" (Singer 1994, 182). Meanwhile, they may not be considered as such. The ability to suffer, which should elicit 'concern,' does not of itself confer personhood or admit a nonhuman animal or animal species to the 'community of equals.' Even to be a nonhuman 'person' on the highest level, within this universe of thought, is to be a poor contender according to its standards of value: the vaunted chimpanzees rank with "intellectually disabled human beings," in Singer's view (Singer 1994, 183).

In the 2011 edition of his book *Practical Ethics*, certain other animals, including some wild birds, are said to perhaps be eligible to be granted some degree of personhood based on laboratory experiments and field observations suggesting that they possess a measure of 'rationality,' 'self-awareness' and future-directed thinking and desires. However, a sentient 'nonperson' or 'merely conscious' being does not qualify for what Singer, citing contemporary American philosopher Michael Tooley, calls a "right to life, in the full sense" (Singer 2011, 85).

I argue that parsing the cognitive capabilities of nonhuman animals in this way relegates the entire animal kingdom, apart from humans, to a condition of mental disability that is totally incompatible with the cognitive demands exacted upon real animals in the real world. It illogically implies a cerebral and experiential equivalence between the intellectually disabled members of one species and the mentally competent members of other species. Rather than helping animals, this model is more likely to hinder the effort, since most people are not likely to care very much what happens to creatures whom even the animal protection community characterizes as mentally inferior and 'disabled.' Ranking animals according to a cognitive scale of intelligence is an aspect of cross-species comparisons that should be avoided.

I first expressed my concern about ranking animals in *Between the Species: A Journal of Ethics* (Davis 1988). In "The Otherness of Animals," I asked whether dogs and cats could be adversely affected if science (or 'science') should decide that they are not as smart as pigs and porpoises. I thought about the dogs I grew up with, and about my Blue-fronted Amazon parrot Tikhon, who, I was told by a bird rehabilitator in San Francisco in the 1970s, was not 'really' intelligent, but a creature of mere 'instinct,' and thus a kind of imposter who only seemed to be an intelligent, emotional and reciprocal companion of mine. In this view, I was a sort of dupe who could not distinguish fixed behavior patterns from conscious awareness in a bird whose ability to fool me depended on the fact that I loved her and needed to believe that we were bonded.

In short, I wanted Tikhon to be intelligent; therefore she was. And since most people do not want chickens and other animals they eat to be intelligent, therefore they are not. This being so, we need to consider, for example, whether we are helping 'food' animals by elevating pigs above chickens, cows and other animals in the food producing sector by making pigs the, as it were, 'great apes' of the farmed animal advocacy project, as in Singer's assertion that of all the animals currently eaten in the Western world, "the pig is without doubt the most intelligent," endowed with an intelligence that is "comparable and perhaps even superior to that of a dog" (Singer 1990, 119). But what do we really know about the total mental capabilities of any animal that is so conclusive that we can confidently state, without doubt, that this one or that one is the most, or the least, intelligent? I would also ask what good it does to tell people that their companion dog may not be as smart as a pig, which raises the issue of pitting animals against one another, as if animal advocacy were an IQ contest of winners and losers.

Can science help us surmount our prejudicial attitudes toward nonhuman animals in order to attain a more just understanding of who they are in themselves, bearing in mind that 'they' are not a monolithic entity ascending through Nature like the floors of a skyscraper from bottom to top?

Not long ago it was generally assumed 'without doubt' that birds were mentally inferior to mammals. Twentieth-century studies upset this assumption. Among birds, in addition to Konrad Lorenz's pioneering studies of geese, jackdaws, and other birds he knew personally and wrote about, pigeons attracted significant scientific interest in the twentieth century due to their homing abilities and their use as messengers in war. Pigeons demonstrate an astonishing ability to handle complex geometrical, spatial, sequential, and photographic concepts and impressions, to solve all kinds of complicated problems, retain precise memories, and invent ways to communicate their understanding, intentions, and needs to human beings. In *Minds of Their Own: Thinking and Awareness in Animals*, Lesley J. Rogers summarizes pigeons' conceptual feats in tests that I personally would fail. Yet despite the evidence, Rogers cites a situation in which a scientist who demonstrated complex cognition in pigeons, including self-awareness, perversely insisted that "if a bird can do it, it cannot be complex behaviour and it cannot indicate self-awareness of any sort" (Rogers 1997, 30, 66–69, 72).

More recently, science investigator Irene Pepperberg, who held firm in a frequently hostile environment of skepticism toward her work, highlighted the intelligence of parrots, based on her years of laboratory experiments designed to coax certain cognitive responses from her African Gray parrot Alex, from the correct use of human verbal language to complex discriminations among shapes, colors, objects, and relationships (NOVA scienceNOW 2011). It may be assumed that these experiments, conducted mostly in windowless basements, and in which Alex was treated more like a kindergarten child than an adult individual, barely hinted at Alex's true range and specific nature of intelligence, but one hopes that they opened a door.

Current evidence suggests much more than merely that some birds display signs of intelligence. Parrots, pigeons, crows, wrens, woodpeckers, kingfishers, finches, seabirds, and other birds are now being acclaimed for their hitherto underestimated cognitive capabilities. For instance, it used to be claimed that birds could respond only to the immediate moment, without any sense of before and after. But as Alexander F. Skutch shows with many examples in his book *The Minds of Birds*, "Birds are aware of more than immediately present stimuli; they remember the past and anticipate the future" (Skutch 1996, 13).

In particular, the ground-nesting birds known as galliforms ('cock-shaped') were traditionally denigrated by Western science as stupid 'in spite of their fine feathers.' Chickens, turkeys, pheasants, quails, peafowl, guinea fowl, and a host of other birds believed to have a common ancestor were dismissed without further ado as "unquestionably low in the scale of avian evolution" (Schorger 1966, 70). Among avian scientists, this assumption has been tossed. As bird specialist Lesley J. Rogers writes in *The Development of Brain and Behaviour in the Chicken*, the information obtained from the research she cites in her book "is beginning to change our attitudes to avian species, including the chicken." She says that with increased knowledge of the behavior and cognitive abilities of the chicken has come "the realization that the chicken is not an inferior species to be treated merely as a food source," and that "it is now clear that birds have cognitive capacities equivalent to those of mammals, even primates" (Rogers 1995, 213, 217).

This claim is upheld by The Avian Brain Nomenclature Consortium, an international group of scientists whose paper, "Avian Brains and a New Understanding of Vertebrate Brain Evolution," published in *Nature Neuroscience Reviews* in 2005, calls for a new vocabulary to describe the various parts of a bird's brain, based on the now overwhelming evidence that the bulk of a bird's brain is not, as was once thought, mere 'basal ganglia' coordinating instincts, but an intricately developed organ of intelligence that processes information similar to the way in which the human cerebral cortex operates (The Avian Brain 2005).

Other studies confirm that the avian brain is a complex organ comprising high-level cognition comparable to the cognition of mammals. For example, an article in *Science Daily* in 2013 states that birds possess a range of skills including "a capacity for complex social reasoning" and problem solving. Professor Murray Shanahan, a researcher from Imperial College London, explains that even though birds have been evolving separately from mammals for around 300 million years, they are "remarkably intelligent in a similar way to mammals such as humans and monkeys" (Imperial College London 2013). In "The Chicken Challenge," Carolynn L. Smith and Jane Johnson present the science showing that chickens "demonstrate complex cognitive abilities" (Smith and Johnson 2012, 76). They argue:

> The science outlined in this paper challenges common thinking about chickens. Chickens are not mere automata; instead they have been shown to possess sophisticated cognitive abilities. Their communication is not simply reflexive, but is responsive to relevant social and environmental factors. Chickens demonstrate an awareness of themselves as separate from others; can recognize particular individuals and appreciate their standing with respect to those individuals; and show an awareness of the attentional states of their fellow fowl. Further, chickens have been shown to engage in reasoning through performing abstract and social transitive inferences. This growing body of scientific data could inform a rethinking about the treatment of these animals. (Smith and Johnson 2012, 89–90)

Notwithstanding these findings—including proof that chickens possess empathy based on studies showing, for example, that mother hens develop stress upon seeing their chicks exposed to stressful situations (Bekoff 2011)—the privileging of the great apes, along with a very restrictive model of intelligence, continues to skew much of the animal advocacy and academic discourse about animal cognition. This privileging disturbs people who have come to know and care about birds and many other kinds of animals in the course of direct interactions with and careful observations of them conducted in sanctuary settings as well as in formal studies.

In *Minds of Their Own*, Lesley Rogers argues that while The Great Ape Project has raised critical issues, by placing the great apes above all other forms of nonhuman life, we are still saying that "some animals are more equal than others." She asks whether, guided by this cognitive-scale-of-being way of thinking, we are going to grant rights to "only our closest genetic relatives?" She exposes the fallacy of ranking animals according to their alleged intelligence or awareness, both of which attributes, she says, "are impossible to assess on any single criterion" (Rogers 1997, 194). Instead of ranking animals according to a simplistic IQ system, Rogers argues that we would be more accurate and just in our assessments if we recognized that "there are many different 'intelligences,' rather than ranking all species on the same scale of intelligence" (Rogers 1997, 57).

Even for humans, Rogers says there is no evidence to support applying the single term 'intelligence' to a diverse set of activities; likewise, there is no evidence that different species use the same cognitive processes to carry out similar types of behavior. In short, there are no grounds for asserting *without doubt* that one group of animals is smarter than another. Ethologist Marc Bekoff agrees, stating that ranking animals on a cognitive scale and pitting them against each other as to who is smarter and more emotionally developed, or less intelligent and less emotionally developed, is silly and even dangerous, considering how these comparisons can be used to claim that "smarter animals suffer more than supposedly dumber animals" whereby 'dumber' animals may be treated "in all sorts of invasive and abusive ways" (Bekoff 2013).

As Malcolm Gladwell observes in "The Order of Things," in *The New Yorker*, "Rankings are not benign.... Who comes out on top, in any ranking system, is really about who is doing the ranking" (Gladwell 2011, 74–75).

Cognitive ranking also raises the quandary of anatomical diversity among animals. In the 1970s and 1980s, the ability of chimpanzees to use American Sign Language, or Ameslan, was news. If chimpanzees could learn this version of human language, then perhaps chimpanzees had a cognitive advantage over all other nonhuman animals, entitling them and their great ape cousins to a semblance of 'human rights.' Such ideas underlay the founding of The Great Ape Project in 1994.

An important fact about the chimpanzee's ability to use Ameslan, however, is that it depends upon an anatomical feature that resembles one of ours—manual dexterity. Thus, no matter how unique, intelligent, or willing they may be, any creatures with fins, paws, hoofs, claws or tentacles cannot learn to use (even if capable of understanding) Ameslan. Similarly, chimpanzees appear to be physiologically and anatomically unsuited to using (however competent of understanding) human verbal language, which is why researchers switched to Ameslan. But what about animals who for whatever reason cannot, or will not, communicate in (and on!?) our terms? Whose kind of intelligence is not our kind? Whose modes of experience elude us? Must 'illiterate' animals forgo 'human rights'? Must they be condemned for being who they are and how they are made to an eternal status of 'non-personhood'?

Allied with the cognitive ranking of competent nonhuman animals—who is smarter, a lizard or a lion, a penguin or parrot, a chicken or a chimpanzee?—is the habit of comparing cognitively intact nonhuman animals not only with humans living with intellectual disabilities but also with children who are cognitively incompetent due to developmental immaturity. This type of cross-species comparison, in which adult nonhuman animals are infantilized, has attracted some animal advocates as a way of gaining public sympathy and support for nonhuman animals by placing them in the light of clever and cute yet vulnerable human youngsters. Indeed, there was an item on the Internet about a woman who said she hesitated to eat a ham sandwich because she had heard that a pig is as smart as a toddler.

Classifying competent nonhuman animals together with vulnerable humans, in order to gain legal recognition and protection of these animals' rights, which they cannot assert on their own behalf, is a necessary and just undertaking. As G. A. Bradshaw and Monica Engebretson urge in "Parrot Breeding and Keeping: The Impact of Capture and Captivity": "Science dictates that standards and criteria to assess and protect human well-being accurately extend to parrots and other animals" (Bradshaw and Engebretson 2013, 1). On these grounds, they argue that "a single unitary model of welfare and legal protection" would rightly include both human and nonhuman animals.

I agree with this argument, but contend that the effort to classify competent nonhumans with incompetent humans is misguided insofar as it exceeds the goal of equal legal protections for all vulnerable beings to foster the fallacy of an inherent equivalency between these two groups' actual mental development and functioning. Mature, unimpaired nonhuman animals are not tantamount to intellectually disabled and underdeveloped humans. Neither chimpanzees nor any other animals could survive let alone thrive in a complex social and natural environment if they could only think and function like toddlers. Children and intellectually disabled humans do not create and sustain stable societies. Let us ask: what does an intellectually disabled adult human being who cannot live autonomously in human society have in common, neurologically and experientially, with a fully developed adult cockatoo carrying out complex ecological, social and parental responsibilities in her forest home? What does a two-year-old child have in common with a mentally healthy adult horse? As the eighteenth-century philosopher Jeremy Bentham observed, paradigmatically: "a full-grown horse or dog is beyond comparison a more rational, as well as a more conversable animal, than an infant of a day or a week or even a month, old" (Singer 1990, 7).

Having run a sanctuary for chickens for over thirty years, I am sometimes asked if I think the chickens see me as their mother and if I consider them my 'babies.' In fact, I do not regard adult chickens as babies. As I explain in my essay, "The Mental Life of Chickens," I see the ability of chickens to bond with me and be companionable as an extension of their ability to adapt their native intelligence to habitats and human-created environments that stimulate their natural ability to perceive analogies and to fit what they find where they happen to be to the fulfillment of their own desires and needs (Davis 2012, 20).

The inherently social nature of chickens enables them to socialize successfully with a variety of other species and to form bonds of interspecies affection and communication. But they are not humanoids. They are not phylogenetic fetuses awaiting human contact to stimulate their cognitive potential. They are neither failed nor inferior humans. An adult hen raising her chicks does not think like a six-year old. She thinks like a mother hen, in which respect she shares commonality and continuity with all attentive and doting mothers of all species.

Chickens in my experience have a core identity and sense of themselves as chickens. An example is a chick I named Fred, the sole survivor of a classroom hatching project in which embryos were mechanically incubated. Fred was so large, loud and demanding from the moment he set foot in our kitchen, I assumed he'd grow up to be a rooster. He raced up and down the hallway, hopped up on my shoulder, leapt to the top of my head, ran across my back, down my arm and onto the floor when I was at the computer, and was generally what you'd call "pushy," but adorably so. I remember one day putting Fred outdoors in an enclosure with a few adult hens on the ground, and he flew straight up the tree to a branch, peeping loudly, apparently wanting no part of them.

"Fred" grew into a lustrously beautiful black hen whom I renamed Freddaflower. Often we'd sit on the sofa together at night while I watched television or read. Even by herself, Freddaflower liked to perch on the arm of the sofa in front of the TV when it was on, suggesting she liked to be there because it was our special place. She ran up and down the stairs to the second floor as she pleased, and often I would find her in the guestroom standing prettily in front of the full-length mirror preening her feathers and observing herself. She appeared to be fully aware that it was she herself she was looking at in the mirror. I'd say to her, "Look, Freddaflower " that's you! Look how pretty you are!" And she seemed already to know that.

Freddaflower loved for me to hold her and pet her. She demanded to be picked up. She would close her eyes and purr while I stroked her feathers and kissed her face. From time to time, I placed her outside in the chicken yard, and sometimes she ventured out on her own, but she always came back. Eventually I noticed she was returning to me less and less, and for shorter periods. One night she elected to remain in the chicken house with the flock. From then on until she died of ovarian cancer in my arms two years later, Freddaflower expressed her ambivalence of wanting to be with me but also wanting to be with the other hens, to socialize and nest with them and participate in their world and the reliving of ancestral experiences that she carried within herself.

Those of us who recognize that chickens and other nonhuman animals are not mere replicas of inferior humans are accused by animal exploiters of 'anthropomorphism.' In 2004, a professor of agriculture at Dordt College in Sioux Center, Iowa, gave a talk in which he argued that the animal rights movement consists mainly of urbanites with "anthropomorphized visions of animals." Animal rights people, he said, know animals mainly as pets, and having been taught that humans "really are like animals," these people have a sentimentalized view of animals, he said (O'Rourke 2004, 1567).

Granted, animal rights people may be tempted to try to turn their companion animals into duplicates of themselves, surrounded as so many of us are by machines and multilayers of comfort in an entirely humanized, technologized world into which our animal companions—our cats and dogs and pet birds—must fit along with the home furnishings and appliances. Rhetorically, some animal advocates may be tempted to portray all animals on the planet as existing in a kind of Disneyesque framework of utopian harmony outside of any natural ecological order. It is possible for even the most dedicated animal rights advocate to slide unwittingly from sensitivity to sentimentality toward the members of other species, to the point where the identities, needs and desires of other creatures become artificially fused, or confused, with the advocate's own, resulting in a false anthropomorphism of over-zealous "humanization" of both domestic and free-living animals. That said, the majority of activists I have worked with for more than thirty years are passionate about wanting nonhuman animals to be able to live according to their natures and be respected for who they are. The desire to share our lives companionably with certain animals, and to protect all animals from human cruelty as much as possible, is quite different from the desire to separate our species from the rest of the animal kingdom, except in the role of a controlling, subjugating force.

Animal exploiters brandish the term 'anthropomorphism' to silence criticism of their mistreatment of animals. Ever since Darwin's theory of evolution erupted in the nineteenth century (*The Origin of Species* appeared in 1859), 'anthropomorphism' has been used to suppress objections to the cruel and inhumane treatment of animals and to enforce the doctrine of an unbridgeable gap between humans and other animals—except when convenient, as in the use of nonhuman animals as experimental "models" for human diseases, or dressing them in costumes and making them do demeaning tricks for our amusement. The term 'anthropomorphism,' which originally referred to the attribution of human characteristics to a deity, now refers almost entirely to the attribution of consciousness, emotions, and other mental states, once commonly regarded as exclusively or predominantly human, to nonhuman animals.

While sentimentalized anthropomorphism may be a risk for animal advocates, anthropomorphism based on empathy and careful observation is a valid approach to understanding other species. Indeed, we can only see the world "through their eyes" by looking through our own. The imposition of humanized traits and behaviors on other animals for purely selfish purposes, forcing them to behave in ways that are unnatural to the animals themselves, is not the same thing as drawing inferences about the emotions, interests and desires of animals rooted in our common evolutionary heritage.

Humans are linked to other animals through evolution, and communication between many species is commonplace. Reasonable inferences may be drawn regarding such things as an animal's body language, facial expressions, and vocal inflections in situations that produce comparable responses in ourselves. Chickens, for example, have a voice of unmistakable woe or enthusiasm in situations where these responses make sense. Their body language of "curved toward the earth" (drooping) versus "head up, tail up" is similarly interpretable. Behavioral resemblances do not require an exact match. One may consider these resemblances in terms of the common wellspring from which all experience flows, or in the form of a musical analogy, in which the theme of sentience and its innumerable manifestations hark back to the matrix of all sentient forms. Anthropomorphism conceived in these terms makes sense. One may legitimately formulate ideas about other than human animals, their desires, needs, deprivations, and happiness, that the rhetoric of exploitation seeks to discredit. One may proffer a counter rhetoric of animal liberation that is free both of our penchant to rank animals from top to bottom and our penchant to set ourselves, jewel-like, in a sphere above and beyond the natural world.

Ranking animals according to a cognitive scale of mental and emotional development risks making excuses to violate any animals that scientists wish to tinker with, not only the supposedly 'lesser' species, but also those regarded as 'higher up' yet inferior to humans in their genetic endowment. At the 2013 *Personhood Beyond the Human* conference at Yale University, some presenters suggested that scientists might 'engineer' animals genetically to be more intelligent than they already are, while others suggested that certain technological inventions of ours—the artificial intelligences—might eventually qualify for moral considerateness and even the status of 'personhood.' Considering that we know almost nothing about the ways in which other animals' intelligences relate to the totality of their being, including their own well-being and sense of self, and considering that we are nowhere near to granting legal or moral considerateness or even a modicum of compassionate treatment, let alone 'personhood,' to billions of sensitive and intelligent birds and other creatures suffering in laboratories and factory farms, these prospects prompt a legitimate concern (Davis, 2013).

In that an animal's brain is an integral part of an animal's body, the idea of genetically engineering other animals' brains to 'enhance' their cognitive capacities seems more like anthropomorphic arrogance than an advancement of ethics or empathy. The idea contradicts and subverts the Nonhuman Rights Project's goal of obtaining legal recognition and protection of an animal's fundamental right of bodily integrity and liberty.

The notion of a brain disconnected from the animal in whom it is situated is implicit in proposals to 'enhance' the mental capabilities of other creatures via surgical or genetic manipulation. In "Brains, Bodies, and Minds: Against a Hierarchy of Animal Faculties," David Dillard-Wright rejects the 'decapitation' theory of consciousness as "a static entity or essence in-residence," observing, rather, the intricate processes and intelligences of the body and the continuity of body and brain, the brain itself being a body part as much as our blood, lungs and kidneys are (Dillard-Wright 2012, 204). Biological situation of brains within and as constituents of bodies, which are themselves environmentally situated and interactive with their surroundings, integrates with all of the evidence we have of evolutionary continuity among animal species and a reasoned belief that other animals' minds are not mere 'precursors' of human ways of knowing but 'parallel' ways of being mentally active and alive in the world (Dillard-Wright 2012, 207).

It might seem that proposals to enhance the cognition of nonhuman animals are in opposition to proposals to expunge their cognition in order to fit them 'more humanely' into our abusive systems. Philosopher Peter Singer, agribusiness philosopher Paul Thompson, and architecture student Andre Ford are among those who have variously supported 'welfare' measures which they claim would reduce the suffering of industrially-raised chickens by inflicting injuries that include de-winging, debeaking, blinding, and de-braining them (Broudy 2006, Thompson 2007, Solon 2012). Asked if he would consider it ethical to engineer a 'brainless bird, grown strictly for its meat,' Singer said it would be 'an ethical improvement on the present system, because it would eliminate the suffering that these birds are feeling.' Ford touted what he called the 'headless chicken solution'—excision of the cerebral cortex—to the suffering of factory-farmed chickens. Proposals to enhance or expunge animal consciousness actually have much in common. Both proceed from presumptions of human entitlement to reconfigure the bodies and psyches of other creatures to fit our schemes and satisfy our lust for manipulating life to reflect our will. Both involve rationalizations that the animals targeted for these procedures are not victims but beneficiaries of the suffering (the injury, wound, harm, trauma) that our species sees fit to impose on them 'for their own good.'

In fact, genetic removal of the avian brain, far from 'removing suffering,' takes suffering to its ultimate limit by destroying the birds themselves, exterminating the identity and *beingness* of birds, and replacing who they are with the anthropomorphized identity of their destroyer. De-braining chickens simply takes the dominant poultry production systems, and the pathologies they incorporate, to their logical conclusion.

It is not unreasonable to worry that robots could be granted a status of legal and ethical 'personhood' long before, if ever, chickens and the majority of nonhuman animals are so elevated. The problem includes but goes beyond the quandary of nonhuman animal diversity in anatomy and physiology mentioned earlier. The minds and personalities of chickens, chimpanzees, and other nonhuman animals will never be able to compete against the dazzle of computers and digital wonders that intoxicate so many of the kinds of people whose power and ambition are charting the course of the planet. How can nonhuman animals, whose intelligences however 'high' are deemed inferior to ours, even by many of their so-called defenders, compete with machines that so many enthusiasts tout as even 'smarter' than we are?

At the same time as these worries loom over nonhuman animals, there are signs pointing in a different direction that could lead to a different conclusion. In "According Animals Dignity," published in *The New York Times*, Op-Ed columnist Frank Bruni draws attention to what he sees as "a broadening, deepening concern about animals that's no longer sufficiently captured by the phrase 'animal welfare'" (Bruni 2014). Citing examples, including the Nonhuman Rights Project, Bruni argues that we are entering an era of "animal dignity" in modern society. The signs of this era, he says, are "everywhere." The attribution of dignity to nonhuman animals by a respected writer in a prestigious, internationally read newspaper is encouraging. It is one of the promising signs of which Bruni speaks, and I hope that his words are prophetic.

References

Bekoff, Marc. "Are Pigs as Smart as Dogs and Does it Really Matter?" *Psychology Today*, July 29, 2013. Retrieved from http://www.psychologytoday.com/blog/animal-emotions/201307/are-pigs-smart-dogs-and-does-it-really-matter.

Bekoff, Marc. "Empathic Chickens and Cooperative Elephants: Emotional Intelligence Expands its Range Again." *Psychology Today*, March 9, 2011. Retrieved from http://

www.psychologytoday.com/blog/animal-emotions/201103/empathic-chickens-and-cooperative-elephants-emotional-intelligence-expan.

Bradshaw, G. A., and Monica Engebretson. "Parrot Breeding and Keeping: The Impact of Capture and Captivity." Animals and Society Institute Policy Paper, 2013. Retrieved from http://www.upc-online.org/thinking/parrot_captivity_impact.pdf.

Broudy, Oliver. "The Practical Ethicist." *Salon*, May 8, 2006. http://www.salon.com/2006/05/08/singer_4.

Bruni, Frank. "According Animals Dignity." *The New York Times*, January 14, 2014. A27. http://www.nytimes.com/2014/01/14/opinion/bruni-according-animals-dignity.html.

Cavalieri, Paola, and Peter Singer (Eds). *The Great Ape Project: Equality Beyond Humanity*. New York: St Martin's Griffin, 1993.

Davis, Karen. "The Provocative Elitism of 'Personhood' for Nonhuman Creatures in Animal Advocacy Parlance and Polemics." Paper presented at the *Personhood Beyond the Human* conference, New Haven, CT, Yale University. December 8, 2013. Retrieved from http://www.youtube.com/watch?v=tQU6ouvs2F0.

Davis, Karen. "The Mental Life of Chickens as Observed Through Their Social Relationships." In *Experiencing Animal Minds: An Anthology of Animal-Human Encounters*, edited by J. A. Smith and R. W. Mitchell, 13–29. New York: Columbia University Press, 2012. Retrieved from http://www.upc-online.org/thinking/social_life_of_chickens.html.

Davis, Karen. "The Otherness of Animals." *Between the Species: A Journal of Ethics* 4.4 (1988): 261–262.

Dillard-Wright, David. "Brains, Bodies and Minds: Against a Hierarchy of Animal Faculties." In *Experiencing Animal Minds: An Anthology of Animal-Human Encounters*, edited by J. A. Smith and R. W. Mitchell, 201–216. New York: Columbia University Press, 2012.

Gladwell, Malcolm. "The Order of Things." *The New Yorker*, February 14, 2011. Retrieved from http://www.newyorker.com/reporting/2011/02/14/110214fa_fact_gladwell?currentPage=1

Imperial College London. "Birds and Humans Have Similar Brain Wiring." *Science Daily*. July 17, 2013. Retrieved August 6, 2013, from http://www.sciencedaily.com/releases/2013/07/130717095336.htm.

Institute for Ethics and Emerging Technologies et al. *Personhood Beyond the Human* conference. New Haven, CT: Yale University, December 6–8, 2013. Retrieved from http://nonhumanrights.net.

NOVA scienceNOW. Profile: Irene Pepperberg & Alex. *Public Broadcasting Service*, February 9, 2011. Retrieved from http://www.pbs.org/wgbh/nova/nature/profile-irene-pepperberg-alex.html.

O'Rourke, Kate. "The Animal Rights Struggle." *Journal of the American Veterinary Association*, May 15, 2004, 1567–1568.

Rogers, Lesley J. *Minds of Their Own: Thinking and Awareness in Animals*. NSW: Allen & Unwin, 1997.

Rogers, Lesley J. *The Development of Brain and Behaviour in the Chicken*. Oxon, UK: Cab International, 1995.

Schorger, A. W. *The Wild Turkey: Its History and Domestication*. Norman: University of Oklahoma Press, 1966.

Singer, Peter. *Practical Ethics*. New York: Cambridge University Press, 2011.

Singer, Peter. *Rethinking Life and Death: The Collapse of Our Traditional Ethics*. New York: St. Martin's Press, 1994.

Singer, Peter. *Animal Liberation*. New York: Avon Books, 1990.

Skutch, Alexander F. *The Minds of Birds*. College Station: Texas A&M University Press, 1996.

Smith, Carolynn L, and Jane Johnson. "The Chicken Challenge: What Contemporary Studies of Fowl Mean for Science and Ethics." *Between the Species* 15.1 (2012): 75–102. Retrieved from http://digitalcommons.calpoly.edu/cgi/viewcontent.cgi?article=2005&context=bts.

Solon, Olivia. "Food Project Proposes Matrix-style Vertical Chicken Farms." *Wired*, February 15, 2012. Retrieved from http://www.wired.co.uk/news/archive/2012-02/15/andre-ford-chicken-farming.

The Avian Brain Nomenclature Consortium. "Avian Brains and a New Understanding of Vertebrate Brain Evolution." *Nature Neuroscience Reviews* 6 (February 2005): 151–159. Retrieved from http://www.avianbrain.org/papers/avianbrainnomenclature.pdf.

Thompson, Paul. "Welfare as an Ethical Issue: Are Blind Chickens the Answer?" *Bioethics Symposium: Proactive Approaches to Controversial Welfare and Ethical Concerns in Poultry Science*, U.S. Department of Agriculture/CSREES/PAS: 3–5, 2007.

Wise, Steve. *Nonhuman Rights Project*, 2014. Retrieved from http://www.nonhumanrightsproject.org.

CHAPTER 11

"I Need Fish Fingers and Custard": The Irruption and Suppression of Vegan Ethics in *Doctor Who*

Matthew Cole and Kate Stewart

Doctor Who, produced by the BBC (British Broadcasting Corporation), is the world's longest running television science fiction series, celebrating its 50th anniversary in 2013. The original 'classic' run of *Doctor Who* was broadcast in the UK between 1963–1989, before going on hiatus, excepting a 1996 TV movie, although the programme retained a cultural presence through the production of new stories in alternative media formats and in video and DVD releases of the 'classic' series. The television series was rebooted in 2005 and has been in continuous production since. Both classic and new series *Doctor Who* targets a 'family' audience, thereby engaging children and their caregivers in a communal viewing experience. This is reflected in the Saturday tea-time (approximately 5pm) scheduling of the original programme in the UK and in its current scheduling early on Saturday evenings (around 7pm). It thereby provides an opportunity for the inter-generational reproduction, or critique, of cultural and ethical norms through the stories it tells and especially the behavior of the central characters and their relations with alien 'others'. The Doctor himself is a member of an alien species, a Timelord from the planet Gallifrey, who has the technological capacity to travel through time and space. The Doctor appears human, with his alien-ness being marked by his technological prowess, longevity, nonhuman anatomy (such as having two hearts) and a range of character quirks that vary with each incarnation. The Doctor is accompanied on his adventures by a changing series of companions, who tend to be human, young, female and white, albeit with more diversity of casting in the post-2005 show. The longevity of the programme is attributable to his capacity to 'regenerate', and therefore be played by a different actor. Thirteen actors have played the part to date in the televised programme: all white British men,[1]

---

1   The first 6 actors to play the Doctor used an RP (received pronunciation) accent, associated with a middle or upper class identity and typical on British television until a more recent trend towards regional diversity on screen. The 7th Doctor, Sylvester McCoy, spoke with a mild Scottish accent, while the 9th Christopher Eccleston, retained his Northern English accent. The current Doctor Peter Capaldi speaks with his native Scottish accent. The age of

despite periodic rumors of a female and/or black iteration of the character.[2]

A common theme within the show is the Doctor saving humankind from exploitation or extermination by alien Others, in which the horror of the story inheres in the objectification of human beings through enslavement, as consumable resources, or as worthless 'vermin'. As such, *Doctor Who* frequently explores a science fiction trope of exploding hubristic human 'superiority' in the face of technologically and/or intellectually superior alien threats, traceable to the roots of the genre: In H. G. Wells' *The War of the Worlds* (2005), first published in 1898, humans are consumed as food by the invading Martians. In the *Doctor Who* episode *The Runaway Bride* (2006), the alien Racnoss declares, "harvest the humans. Reduce them to meat". A variant on this theme is the Doctor's more direct critique of the exercise of human 'superiority' over and against vulnerable alien others. For example, in the 2007 episode *42* (2007), the Doctor lambasts the human crew of a spaceship for their exploitation of a living sun for fuel: "Humans, you grab whatever's nearest and bleed it dry". These narrative devices facilitate the construction of the Doctor as a heroic figure, saving humankind either from an alien peril, or from the consequences of our own moral outrages, over and over again. It thereby facilitates the construction of the Doctor as a moral leader, through his critique of the violent, hubristic and exploitative ambitions of both alien Others and humans ourselves. This moral leadership role is acknowledged as central to *Doctor Who* by Julie Gardner, the Executive Producer of the rebooted show: "Your actions

---

the actor playing the part has varied widely: the two oldest bookend the series to date: Peter Capaldi is the oldest actor to begin working in the role, followed by first Doctor William Hartnell; both in their mid-50s when taking the part. However, the un-numbered 'War Doctor', was played by John Hurt while in his 70s (see note 2). Matt Smith, the 11th Doctor, is the youngest to play the role to date, being cast when aged 26.

2 The numbering of the Doctors has been complicated in recent years by the insertion of an un-numbered 'War Doctor', played by John Hurt, in two televised episodes and one 'webisode' in 2013, into the fictive history of the character. The War Doctor incarnation lived between the 8th and 9th incarnations. For the purposes of this chapter, the War Doctor is discounted in the numbering. Therefore, the numbering of Doctors in the text refers to: 1st (William Hartnell, 1963–1966); 2nd (Patrick Troughton, 1966–1969); 3rd (Jon Pertwee, 1970–1974); 4th (Tom Baker, 1974–1981); 5th (Peter Davison, 1981–1984); 6th (Colin Baker, 1984–1986); 7th (Sylvester McCoy, 1987–1989 and 1996); 8th (Paul McGann, 1996); 9th (Christopher Eccleston, 2005); 10th (David Tenant, 2005–2010); 11th (Matt Smith, 2010–2014); 12th (Peter Capaldi, 2013-date). In addition to their main tenures as the Doctor, most of the actors have reprised the part in later televised 'multi-Doctor' episodes and/or in spin-offs or other media. Several other actors have also played or parodied the part in various media.

count. The universe is dangerous and your moral choices, your actions define you. If we all had a little more of The Doctor in us, our world would be a better, braver place" (Gardner 2009).

Taking our cue from this explicit foregrounding of moral leadership in the rebooted show, in this chapter we explore how televised *Doctor Who* since 2005 deals with a central pillar of hubristic human 'superiority'; the exploitation of other animals for human food, in the euphemized forms of 'meat' and other 'animal products'. While the entire 51 year run of the televised programme and its spin-offs in other programmes and other media are rich sources of data for an exploration of this theme, we largely limit ourselves in this discussion to a focus on the post-2005 televised series, partly as a pragmatic means of delineating our analysis and partly in order to foreground *Doctor Who's* most contemporary and highest profile manifestation in popular culture. We begin with a closer examination of the construction of the Doctor as a moral leader or guide within the show and argue that this construction logically and ethically ought to incline the Doctor towards vegan practice and the espousal of vegan ethics. We follow that with an analysis of the manifest failure of the show to fulfill that promise; the Doctor's love for 'the other' is undercut by his normalization of 'meat' eating and other exploitative practices re: 'real' other animals. This failure is then contextualized by the anthropocentric and speciesist social context of the show, before we end by exploring some exceptions to this illogical exploitative norm within *Doctor Who*. Appropriately but ironically enough, this takes us back in time to the classic era of the programme, and the Doctor's on-screen conversion to vegetarianism in 1985 (a conversion that was subsequently underplayed and deliberately abandoned). This provides an exemplification of the potential of the series to reach the logical conclusions of its own ethical premises. We therefore argue for the restoration and extension of the Doctor's epiphanic 1985 conversion through the future veganization of the character.

Loving the Alien: The Valorized Ethics of the Doctor

The character of the Doctor did not begin as a moral leader. Infamously, the first iteration of the character, played by William Hartnell, appeared to contemplate bludgeoning an injured man who was slowing the escape of the Doctor and his human companions from danger, in the very first broadcast serial, *An Unearthly Child* (1963). The moral ambiguity of Hartnell's initial portrayal of the Doctor was gradually refined into a more straightforwardly heroic character. By the time of the 2005 re-launch, there was no doubt that the Doctor's defining characteristic was heroism; a heroism that inhered in his capacity to provide moral guidance for his fictional human companions and therefore by extension, for his 'family' audience.[3] This interpretation was made explicit in the final episode of the first run of the new series, in which the Doctor's companion, Rose Tyler, declares; "The Doctor *showed me a better way of living your life* [...] That you don't just give up. You don't just let things happen. You make a stand. You say no. You have the guts to do what's right when everyone else just runs away" (*The Parting of the Ways* 2005, emphasis added). The Doctor is thereby defined by doing the right thing, regardless of the personal risk and regardless of any evasion of moral responsibility displayed by others around him. The Doctor is prepared to be an ethical outsider; that is to be different from the crowd not merely by dint of idiosyncratic costume or mannerism (a theme we return to below) but through his ethically motivated actions.

The content of the Doctor's ethics centers on his attitude of wonder, reverence and love for 'the other', which supplants fearful or hostile responses to aliens that are frequently displayed by human characters in the show. This was expressed as the central message of both classic and new *Doctor Who* in an interview by BBC *Breakfast* with Christopher Eccleston, who portrayed the 9th iteration of the Doctor in the 2005 run of 13 episodes; "The classic elements of this series are there...the central message of love life in all its forms...he doesn't react with horror when he sees a blue, three-headed monster, he reacts with wonder and I think that's a very important message to send out to children" (Eccleston 2006). Eccleston's emphasis on wonder plays out repeatedly in the new series, including in that of his successors David Tennant (the 10th Doctor), Matt Smith (the 11th Doctor) and Peter Capaldi (the 12th and current Doctor). Tellingly, an attitude of wonder is expressed even in the face (sometimes literally) of dangerous and hostile others. The 10th Doctor responds to seeing an alien resembling a werewolf with a dreamily spoken, "oh that's beautiful" (*Tooth and Claw* 2006), comments "well, you are wonderful" at a giant wasp (*The Unicorn and the Wasp* 2008) and expresses his sympathy for the alien Isolus, a creature that possesses a young girl in *Fear Her* (2006). Similarly, the 11th Doctor remarks, "Oh look at you, you are beautiful" on encountering a Minotaur-like creature in *The God Complex* (2011, 6/11) and opines, "you are beautiful" on removing the mask of Alaya (a member of the species 'homo reptilia', who was being held captive after attacking the Doctor

---

3 The post-2005 show has somewhat deprivileged the Doctor as a moral leader by, on occasions, portraying his human companions as having greater ethical insight than him. For example, Clara admonished the 12th Doctor for failing to grasp the potential for moral reform of a Dalek; his most famous alien enemy (*Into the Dalek* 2014).

and his companion Rory). The 11th Doctor expresses child-like glee at discovering 'dinosaurs on a spaceship' in the episode of the same name (2012), while the 12th Doctor appears to flirt with an oversized female *Tyrannosaurus Rex* transported to the Thames in Victorian London in his debut episode, *Deep Breath* (2014). By this point in the new series, another member of 'homo reptilia', Madame Vastra, has become established as a recurring character in *Doctor Who*, along with her human wife Jenny and their associate/Butler Strax, a member of the Sontaran species. Strax is played for laughs in the show, despite the Sontarans having been portrayed as remorselessly violent and hostile to humans in their previous appearances, stretching back to the 3rd Doctor series, *The Time Warrior* (1974). Strax therefore highlights the redeemable qualities of otherwise threatening alien others from the Doctor's perspective, subverting the prior tendency to objectify Sontarans as purely monstrous and callous in the show: in the 4th Doctor story *The Sontaran Experiment* (1975), the Sontaran Lynx vivisects unanaesthetized human victims. The Doctor therefore displays moral leadership for the audience through his capacity to judge individual others on their merits, rather than to stereotype and stigmatize them. This is especially significant in the case of Strax, given that Sontarans are established as a cloned warrior species in the show and given their violent past appearances.

These examples of the Doctor's wonder and acceptance of 'otherness', encompass both extra-terrestrial and terrestrial creatures ('homo reptilia' are not extra-terrestrials, but are described by the 11th Doctor as "the previous owners of the planet [Earth]", highlighting a speciesist hierarchizing that we return to below). In our discussion so far they do not include examples of nonhuman animals who are exploited for 'meat' or other human uses in the non-fictional cultures of the audience; that is, the Doctor's expressions of wonder and love for others tends to be focused in such a way as to avoid drawing attention to the manifest withholding of wonder and love from the animals who humans confine and kill en masse in the nonfictional world. There are, however, exceptions to this rule. The 9th Doctor responds to discovering that a pig has been vivisected by a family of aliens invading Earth, the Slitheen, with outrage: "They've taken this animal and turned it into a joke", and cradles her or him in his arms as she or he dies (*Aliens of London* 2005). Despite his moral outrage, it is striking that the 9th Doctor de-genders the pig and asserts her or his general nonhuman animalness over her or his species-specific pigness, with the effect that the implicit positioning of the pig as worthy of moral concern is attenuated. Contrastingly, in the 2010 Christmas episode, *A Christmas Carol*, the 11th Doctor implicitly corrects a little boy who refers to a threatening flying shark as 'it' by referring to her as 'she'. Furthermore, he wonderingly intones, "look at you sweet little fishy-wishies" when he encounters a school of 'fish that can swim in fog' in the same episode. However, this non-violent relationship with fishes stands in stark contrast to the 11th Doctor's consumption of fishes throughout his tenure in the show; Doctors 9, 10 and 11 are all enthusiastic 'meat-eaters'.

Eating the Other: The Doctor's Ethical Lacuna

The reproduction of 'meat'-eating, and other forms of 'animal product' consumption, is ubiquitous in popular culture, to the point of banality. That very banality however, is sociologically problematic given the gross incongruity between the cultural valorization of compassion, care and concern for other animals on the one hand, contrasted with the massive, industrialized violence perpetrated against them on the other. That incongruity can be sociologically understood as a state of denial (Morgan and Cole 2011) that is especially acute in respect of the socialization of children into dominant values and forms of relation with other animals (Cole and Stewart 2014; Stewart and Cole 2009). In this context, the post-2005 series of *Doctor Who* by and large reproduces the banal consumption of animal products as normal, pleasurable and unquestionable, in spite of the show's valorized ethics of compassion, wonder and love for the other discussed above. The banality of 'meat'-eating and its normalization as beyond the realm of ethical questioning are closely related. That is, it is precisely because 'meat'-eating, and other forms of consumption of nonhuman animals' bodies, is so routinized as a taken-for-granted everyday pleasure that ethically questioning the practice can be comfortably elided in *Doctor Who*, despite its illogicality in the context of the Doctor's own ethics. This can be illustrated by the banality of the Doctor's own relations with exploited nonhuman animals in the show.

Christopher Eccleston's performance as the 9th Doctor is marked by his wearing of a cow's skin ('leather') jacket in every episode of his tenure. In his first episode he also requests milk in his cup of tea from his new human companion, Rose (*Rose* 2005). In his third episode, *The Unquiet Dead* (2005), he is again served milky tea and by his ninth episode, *The Empty Child* (2005), Eccleston's Doctor and Rose have bonded to the extent of bantering about the experience of running out of (cow's) milk:

> Doctor: "You know how long you can knock around space without happening to bump into Earth?"
> Rose: "Five days. Or is that when we're out of milk?"

> Doctor (with a look of amused mock astonishment): "Of all the species in all the universe and it has to come out of a cow."

This mild expression of distaste represents a rare case of the *explicit* slippage of the commitment to love for 'the other' in *Doctor Who*, as contrasted with the more normal *implicit* slippage represented by his banal consumption of animal products. For example, later in the same episode, the Doctor helps himself to sliced roasted 'meat' at a meal shared with a group of urchin children, after which he is admonished by the child character Nancy for greedily helping himself to two slices. Slippages such as the earlier allusion to the bizarreness of drinking the milk of other species open up possibilities for ethical critique; possibilities which are generally not taken up in *Doctor Who*, an issue we return to in the conclusion to this chapter.

Although these examples are striking in themselves for the normalization of 'meat' eating and other uses of nonhuman animals in the programme, the clearest breach with the 6th Doctor's 1985 vegetarian conversion comes in the episode *Boom Town* (2005).[4] In a lengthy restaurant scene, the 9th Doctor looks at the menu and declares his choice to the waiter, "Mmm steak looks nice; steak and chips!" (although the action develops so that he is never shown being served or eating the steak). This meal choice of the 9th Doctor is especially significant for three reasons. Firstly, it facilitates the empathetic identification with the Doctor on the part of the audience, because it demonstrates his (pun intended) 'down to earth' character, in spite of his alien-ness; the Doctor is 'like us' (as long as 'us' excludes vegans or vegetarians). This interpretation is reinforced by the following exchange between Rose and her mother, Jackie, in the earlier episode *World War Three* (2005).

> Jackie: "What does he eat?"
> Rose: "How do you mean?"
> Jackie: "I was going to do shepherd's pie.[5] All of us. A proper sit down... I mean I don't know, he's an alien. For all I know, he eats grass and safety pins and things."
> Rose (laughing): "He'll have shepherd's pie."

---

4   "The Doctor himself turns vegetarian in 1985's *The Two Doctors*. A 1986 comic has him lapse; but Paul Cornell's 1995 novel *Human Nature* (adapted for TV in 2007) suggests that he's still vegetarian in his subsequent, seventh, incarnation." (The Vegan Option 2013).

5   Shepherd's pie is a traditional British dish based around minced lamb's flesh and topped with mashed potatoes.

Secondly, it models the successful performance of the Doctor's incongruous ethical stance in relation to exploited others, because it occurs in the context of an extended debate with his Slitheen captive (going by the human name of Margaret) about the ethics of violence. In response to Margaret's defense of herself on the basis of her sparing the life of one potential human victim, the Doctor launches into an impassioned tirade against his captive's ethics that could logically function as a critique of the denial and emotional gymnastics that underpins 'meat'-eating in the non-fictional world:

> You let one of them go but that's nothing new. Every now and then a little victim's spared, because she smiled, because he's got freckles. Because they begged. And that's how you live with yourself. That's how you slaughter millions. Because once in a while, on a whim, if the wind's in the right direction, you happen to be kind. (*Boom Town* 2005)

This account could equally apply to the periodic media celebration of escaped nonhuman animals, who evade the fate of the slaughterhouse and live out their lives in a sanctuary, with their subjecthood at least partially restored through their discursive individualization as media 'characters' (Morgan and Cole 2011). In this light, the Doctor's words could work as a scathing condemnation of the hypocrisy of 'meat' culture at large, in which the Slitheen Margaret stands as a proxy for the monstrousness of humans' treatment of other earthlings who are killed for their flesh. But, this meaning is made less accessible for the audience given its undercutting by the Doctor's own menu choice in the restaurant. Therefore, instead of opening up the difficult question of humans' relations to our sentient 'prey', the exchange forecloses it, demonstrating the social fact that the lives and deaths of most other animals simply do not count when it comes to ethical debate in mainstream non-vegan society.

Thirdly, and by extension, this juxtaposition of anticipated 'meat'-eating and impassioned critique of violence models and thereby normalizes that incongruity in the wider culture and for the specific *Doctor Who* audience: illogical ethics are de-problematized and trumped by the 9th Doctor's taste preference for 'steak', such that the violent actions that 'produced' that 'steak' are removed from the sensibility of the audience. In Carol J. Adams' terms, the cow becomes an 'absent referent' in the scene (Adams 2004). Violence against other animals simply does not count as violence and is therefore beyond the domain of ethics; the results of violence only exist as objects for pleasurable consumption for the new series' Doctors. The extent to which this reflects the mainstream cultural context of *Doctor Who* was revealed in an interview marking the show's 50th anniversary with Russell T. Davies, the author of *Boom*

Town and many other new series' episodes between 2005 and 2010, spanning the tenures of the 9th and 10th Doctors. Davies, a long-standing fan of *Doctor Who*, was also the 'show runner' of the re-launched series and was instrumental in bringing it back to the television screen. In the segment of the interview reproduced below, he makes plain how the 9th Doctor's eating practices are a reproduction of his own.

> Interviewer: "You made him eat steak and chips [...]"
> Davies: "I remember writing that thinking oh he was vegetarian, it was Colin Baker, they made him vegetarian [...] I'm not vegetarian, thinking I don't want to write a vegetarian doctor [...] and I was taking a deep breath thinking right I'm gonna do it I'm gonna have steak and chips [...] because in an episode that's essentially about a liberal debate [...] I thought how awful if he then orders a vegetarian meal, on top of everything else it's gonna look like it was written on the *Guardian* front desk[6] [...] so I remember thinking it's gotta be steak and chips, cos I love steak and chips and it's like that's a good solid earthy meal."

Davies' inscription of 'earthiness' as a characteristic of flesh-centered meals supports our earlier argument that the new series' Doctors' food practices construct him as a 'down to earth' character for the audience, in spite of his alienness. It is also striking that the setting for the new series is overwhelmingly planet Earth and specifically the UK (the 'steak and chips' scene takes place in a restaurant in Cardiff Bay), albeit sometimes in the past or future, contrasting with the more frequent excursions to other worlds or other Earthly cultures in most periods of the 'classic' programme. This theme of figuratively grounding the Doctor in contemporary UK culture is further explored in relation to entrenching the show's populist appeal under Davies' leadership:

> Interviewer: "but that does strike me as a bit of you because you are obviously an intelligent and artistic person, but you are a populist intellectual [...]"
> Davies: "I hope so, thank-you yes."
> Interviewer: "[...] you're not a sneering liberal [...] there is an element of the sort of liberalist arts movement can do itself no favours by being lofty."

---

6 *The Guardian* is a national newspaper in the UK that is generally left-leaning in its politics and socially liberal.

> Davies: "Yes absolutely yes, it would have been awful if he'd ordered alfalfa omelette in that scene [...] and I think a good chunk of the audience would have gone, 'oh you're nothing to do with me now', seriously, they would've gone, 'no thanks' [...] lines like 'steak and chips', they are important in the end [...]for the past eight years you can see the programme keep realizing that this man is real, this Doctor who travels in a TARDIS and is a Timelord is really really real.'
> [...]
> Davies: "Obviously he's not human, he's a Timelord but he eats [...] he has steak and chips, he's hungry [...] and when he enters a story he really feels this, he doesn't just walk through the door and go hello, actually he's fascinated and he becomes part of the story and if people die it hurts and if something's funny it's hilarious, you know? It's it's he's real. It is a significant difference between the old and the new series [...]" (2014).

Davies' connection of 'meat'-eating with being 'real' reproduces an elitist stereotype about non-meat eaters that is compounded by the interviewer's interjections. Read as a whole, the conversation links vegetarianism with 'sneering liberalism', a lack of emotional feeling and ultimately a lack of ethical commitment, as if eating 'meat' were a key element of the Doctor's moral character. This inverted classism re: vegetarianism/veganism is revealing of Davies' unexamined prejudices but more importantly reproduces a damaging cultural stereotype that equates the refusal of 'meat' with aloofness instead of its reverse; a principled opposition to violence and exploitation towards other animals—the very kind of ethical action that the Doctor pursues in other contexts within the show. Therefore, in spite of the undoubted, albeit uneven, progress that the programme made under Davies' stewardship in respect of tackling the legacy of sexist, racist or heteronormative stereotypes in the classic series (see Orthia 2013), in relation to other animals the new iteration of the show is reactionary, conservative and apparently oblivious to its own reproduction of violent real world relations between humans and other animals.

That obliviousness is further evidenced in David Tennant's portrayal of the 10th Doctor, a period which is peppered with throwaway references to, and the consumption of, 'meat' and other animal products, which demonstrate their banality but that also deepen their normalization for the audience. In his first episode, *The Christmas Invasion* (2005), the Doctor is revived from his post-regeneration torpor by the aroma of a milky cup of tea. Later in the episode he castigates the British Prime Minister for shooting down a retreating alien spaceship; "That was murder". He goes on to characterize humans as the real monsters in light of the Prime Minister's violent action, which she euphemizes as "defense", but the episode concludes with a traditional Christmas dinner scene with carved pieces of a turkey's flesh on the Doctor's plate. In his third episode, *Tooth and Claw* (2006), featuring a werewolf-like alien threatening to possess Queen Victoria, the 10th Doctor requests of his host to "save her a wee bit of ham" in reference to Rose, thereby normalizing 'meat'-eating for his human companion and not just for himself. In his sixth episode, *The Age of Steel* (2006), the Doctor asks Mrs. Moore, a human character carrying a bag of equipment: "Haven't got a hotdog in there have you? I'm starving", to which she replies, "of all the things to wish for. That's mechanically recovered meat." Rather than taking that as a cue to regret his whim, the Doctor's rejoinder is, "I know. It's the Cyberman[7] of food, but it's tasty." As with the 9th Doctor's expression of mild distaste for the provenance of cow's milk, this slippage into dialogue of the usually hidden production process of animal products (the partial restoration of the absent referent) is undercut by the trump card of banal taste preference and any ethical considerations are thereby elided. In his tenth episode, *Love and Monsters* (2006), the 10th Doctor's normalization of 'meat' extends to its comic use as bait for an unnamed alien monster: "who'd like a porky choppy then". Animal products are again discursively deployed for comic effect in the 2007 episode *Blink*, in which the Doctor jokes about his 'timey-wimey' gadget that "can boil an egg at thirty paces, whether you want it to or not actually, so I've learnt to stay away from hens. It's not pretty when they blow". The 'accidental' killing of other animals here is a matter for laughter and merely aesthetic regret, which only makes sense in a cultural context in which hens are already objectified, but which makes no sense in the context of the Doctor's own purported ethics of compassion. Similarly, Tennant's Doctor accepts an offer for a 'buffalo wing' from a passenger on board a spaceship version of the Titanic in *Voyage of the Damned* (2007) and grins without comment at his dinner companion's musing that, "They must be enormous, these buffalo, so many wings". In *Silence in the Library* (2008), he accepts a packed lunch from the character River Song that comprises, "chicken and a bit of salad" and uses a "chicken leg" as bait for the carnivorous Vashta Nerada in the sequel episode.

The latter is one of many examples in *Doctor Who* of the science fiction trope of positioning humans as 'meat' for an alien species, but the irony of the ubiquitous use of the chicken's leg as food goes unremarked on in the episode. The following episode, *Midnight* (2008), involves the Doctor consuming 'meat' again, in the form of an in-transit meal on a tourist vehicle on the titular planet.

---

7   The Cybermen, humanoid cyborgs, are a recurring enemy of the Doctor.

His neighbor in the vehicle asks him, "Well what's this? Chicken or beef?", to which the Doctor replies, "I think it's both". Again, the character is depicted as unconcerned with the ethics of using other animals for flesh, extending to this projected future of an implied hybridization of different species of other animals to satisfy human taste preferences. In an echo of his earlier castigation of the UK Prime Minister for shooting down the retreating alien invaders, Tennant's Doctor later, and without irony, scolds his fellow human passengers for contemplating the murder of an apparently possessed passenger, presumably with the fictive taste of chicken/beef still in his mouth. In a variation on this irony-free normalization of 'meat' culture, in the later 10th Doctor episode *Planet of the Dead* (2009), the Doctor comforts Lou and Carmen, frightened passengers of a bus transported to the titular planet, by inviting them to focus their thoughts on their anticipated evening meal:

> Doctor: "What's for tea?"
> Lou: "Chops. Nice couple of chops and gravy. Nothing special."
> Doctor: "Oh that's special Lou. That is so special. Chops and gravy. Mmm!"

This scene takes place in the context of the passengers being potential prey of a voracious alien swarm on the planet. The theme persists in the next episode (*The Waters of Mars* 2009), in which the Doctor tries to save the human crew of a Martian colony. The following dialogue again illustrates the Doctor's acceptance and approval of mainstream 'meat' culture, even when any pretence at 'meat'-as-necessity is stripped away by the fictive context:

> Doctor: "And you're growing veg!"
> Adelaide: "It's that lot [the rest of the crew]. They're already planning Christmas dinner. Last year it was dehydrated protein. This year they want the real thing."
> Doctor: "Still, fair enough. Christmas."

The implied killing of a turkey (the traditional victim of choice in a British 'Christmas dinner') is not only accepted here, but projected into the future as a continued yearned-for tradition.

The normalization of 'meat'-eating continued into the 11th Doctor's tenure, under the aegis of the then new and at the time of writing still current show runner, Steven Moffat. Moffat had already authored several episodes in Davies' era of the show, with his contributions including Eccleston's throwaway line about cows' milk and Tennant's chicken salad packed lunch discussed above. One of the distinctive traits of Matt Smith's 11th Doctor however, was his more self-evident alien-ness vis-à-vis human beings, through idiosyncratic mannerisms, turns of phrase and mild social faux pas. This included his food practices, but crucially not to the extent of problematizing 'meat' culture. In his first episode, Smith's Doctor tries and vehemently rejects several food items (including apples, yoghurt, bacon, baked beans, bread and butter and carrots), for comic effect, before settling on a 'quirky' combination that would recur throughout his tenure: "I need fish fingers[8] and custard" (*The Eleventh Hour* 2010). The evoked memory of fish fingers and custard later proves to be enough to revive a dying Doctor in *Let's Kill Hitler* (2011). The 11th Doctor's relationship with food continued to be expressed as 'quirky' throughout his four seasons in the role. In *The Hungry Earth* (2010), he regrets that, "I can't make a decent meringue". In *The Lodger* (2010), he drinks cows' milk straight from a bottle (as he does in the later episodes *Closing Time* (2011) and *Hide* (2013) and cooks an omelette for his prospective landlord Craig, cracking hens' eggs, grating cheese, throwing in slices of pigs' flesh ('ham') and squirting in salad cream (a condiment incorporating hens' eggs). As noted above, the 11th Doctor reacts with wonder to the flying fishes in *A Christmas Carol* (2010), but later in the same episode he sits down to share a traditional Christmas dinner with a human family, with a roasted turkey's corpse as its centerpiece. The dinner also follows his harnessing of a flying shark to haul an airborne sled, as a festive joyride. When asked by his sentient time machine, the TARDIS, about fish fingers in *The Doctor's Wife* (2011), the Doctor echoes the 10th Doctor's processed flesh-food preferences by describing them as, "not raw, lovely and cooked, processed food, mmm, fish fingers". The TARDIS responds, "do fish have fingers?", but the Doctor ignores the question, missing the opportunity to make visible the absent referent for himself, and for the audience.

The next story, *The Rebel Flesh* (2011), opens with the 11th Doctor asking his companions Amy and Rory, "who wants fish and chips?", having apparently forgotten any lingering compassion for fishes that he felt previously. This opening gambit of the Doctor's is especially incongruous given that this episode and its sequel, *The Almost People* (2011), contains a powerful critique of the human capacity to mete out violence on the basis of an 'othering' process. The story revolves around a violent dispute between a group of humans and their 'gangers'—replicas of the humans synthetically created to undertake (and be sacrificed performing) dangerous tasks. The story is resonant of two 10th Doctor stories featuring the 'Ood', who are first introduced as a species

---

8 'Fish fingers' are frozen, breaded finger-shaped portions of fishes' flesh which are marketed as a food primarily for children.

'born to serve' (*The Impossible Planet* 2006) and are sacrificed by the Doctor when he saves their human masters. In the second Ood story (*Planet of the Ood* 2008), we learn they have been enslaved by humans who farm them for profit, and the story sees the Doctor and his companion Donna rescuing them from human use. Both the gangers and the Ood invite comparisons with the commodification of nonhuman animals for profit, but the Doctor's ethic of 'love life in all its forms' is undercut by the Doctor's offer of "fish and chips" and by the (unremarked) fur trimmed coat worn by Donna in *Planet of the Ood*.

Fishes recur as food-objects in *Night Terrors* (2011), in which the Doctor asks, "how do you feel about kippers?", playfully flapping one about before cooking them for breakfast for a family of a human couple and an alien 'cuckoo' child with a human appearance. The story is ironic given the Doctor's successful efforts to reconcile the human couple to adopting and loving this alien; the ethics of compassion do not extend to the fishes killed and smoked to produce 'kippers'. The irony continues in *The God Complex* (2011); as mentioned above, the Doctor admires the threatening Minotaur-like creature as 'beautiful', but later in the episode claims to have a degree in cheese-making: the resemblance between a 'beautiful' Minotaur and the cows exploited to 'make cheese' appears to be lost on the character. The 11th Doctor's fondness for dairy products is further developed in his implied claim to have invented Yorkshire pudding in *The Power of Three* (2012); a traditional British dish based on batter made from cow's milk, hen's eggs and flour. In one of his final episodes, *Nightmare in Silver* (2013), he offers "free ice cream" to his companions. Returning to the 11th Doctor's predilection for eating fishes, in a cozy scene on the lounge sofa in *The Power of Three* (2012), the Doctor shares his favorite meal of fish fingers and custard with his companions Amy and Rory, while in his final episode, *The Time of the Doctor* (2013), fish fingers and custard constitute the Doctor's final meal before his regeneration, bookending Matt Smith's tenure in the role. His final episode is also noteworthy for the use of a 'Christmas turkey' as a plot device. The Doctor uses the TARDIS to help his companion, Clara, cook the bird for her family's Christmas dinner. At the start of the episode, the Doctor asks Clara how the turkey is doing, to which she jokes, "dead and decapitated, but that's Christmas when you're a turkey". After transferring the turkey to the TARDIS, the Doctor in turn jokes that, "it'll either come up a treat or just possibly lay some eggs" (as a result of the time travelling capacities of the TARDIS). Later in the episode the resurrection joke is echoed when the Doctor says that the turkey is finally cooked; "either that or it's just woken up". Removing the cooked turkey from the TARDIS's inner workings, Clara asserts that the "turkey smells good", to which the Doctor agrees; "yeah, smells great". In this example, the Doctor's previous critique of the de-gendering of 'the other' in *The Rebel Flesh* has been forgotten. Instead he reproduces that de-gendering himself, in spite of the implicit feminization of the dead turkey in his joke about egg-laying.

From cow's skin as fashion item, through pining for steak, hotdogs and fish fingers, to objectifying jokes about turkeys' destiny as Christmas dinner, the new series' incarnations of the Doctor have remorselessly reproduced the normality of human uses of and violence towards other animals, in spite of an explicitly trumpeted ethics that would logically preclude these actions. The next section addresses how that uneasy combination is able to be sustained within the programme, but also how it sometimes breaks down and the potential that this suggests for reconciling the Doctor's *professed* ethics and his ethical *actions* in respect of nonhuman earthlings.

Selective Compassion: The Tricky Business of Sustaining 'Meat' Culture in *Doctor Who*

In the previous section, we detailed how the new series' Doctor has consistently fallen short of his own ideal of compassion in his behavior towards other animals, and especially in respect of his blasé consumption of animal products in the show. An explanation, but not an excuse, for this illogicality is the anthropocentric and speciesist social context of the programme and its writers and producers. More particularly, this context encompasses both the valorization of human beings and the way that nonhuman animals are selectively afforded subjecthood and cultural sensibility depending on the uses that humans make of them (Cole and Stewart 2014; Peggs 2012). In this section, we explore these inter-related aspects of the social context of the show, but also illustrate how that context is sometimes challenged.

It is notable that the Doctor performs an ethics of compassion only in respect of particular 'types' of others: either those who exhibit specific similarities to human beings, especially through tacit, arbitrary and anthropocentric evaluations of sentience (such as human-like language use or technological prowess), or those who are in some way unusual, spectacular, or exotic. The paradigm example of the former is the Doctor's respect for homo reptilia, while the latter are exemplified by the 'sweet little fishy-wishies' in *A Christmas Carol*, who are spared the Doctor's more usually voracious relationship with fishes, apparently due to their capacity to 'fly' in the fog. In contrast, other terrestrial animals are usually exiled beyond the scope of the Doctor's performative ethics. However, there are exceptions to this norm, where the nonfictional exploitation of other animals seeps through into *Doctor Who* and becomes at least dimly discernible as an ethical issue that ought to concern the Doctor, his companions and the audience.

The 10th Doctor's second episode incorporates a critique of vivisection that echoes the 9th Doctor's outrage at the Slitheen's treatment of a pig (discussed above): In *New Earth* (2006), the Doctor discusses the fate of living humans being used to manufacture treatments for the wealthy clientele of a futuristic hospital: "They were born sick, they're meant to be sick. They exist to be sick. Lab rats. No wonder the sisters [the cat-like staff of the hospital] have got a cure for everything; they've built the ultimate research laboratory; a human farm". The fulcrum of the critique is unclear here: is the Doctor's moral outrage rooted in the analogous outrageousness of nonhuman vivisection as such, or is it rooted in a speciesist complaint against the 'reduction' of human beings to an inferior, subhuman, level? While his outrage demonstrates the possibilities that the show contains for critiquing, or encouraging critique, of nonfictional violence towards other animals, those possibilities remain stifled in the context of the illogicality of the Doctor's wider non-veganism, and his speciesist bias towards human beings over and above other species, especially nonhuman earthlings. A rare example of the Doctor engaging, albeit confusedly, with nonhuman earthlings as ethical subjects occurs in the 11th Doctor episode *A Town Called Mercy* (2012). Set in a fictive frontier town familiar from the Western film genre, the Doctor asks to borrow the horse of a citizen of Mercy, thereby acquiescing to the owner-property relation between humans and horses, but before he rides off, he claims that, "I speak horse" and further asserts that, "he's called Susan and he wants you to respect his life choices". This comedic scene chimes with the post-2005 shows' theme of destabilizing heteronormative gender identities, but it also establishes the Doctor's capacity to communicate with nonhuman earthlings and also establishes their status as agential characters within the show. This is a relatively 'safe' gambit vis-à-vis horses, given their status as non-food animals in British culture, emphasized by the scandalized media and political reaction to the presence of 'horsemeat' in the UK food chain in 2012 (BBC 2013b; 2013c; see also Nik Taylor and Jordan McKenzie's chapter in this volume). However, given the truth of the Doctor's ability to speak with and defend the interests of nonhuman earthlings in this example, his continued consumption of animal products looks even more indefensible in the context of his wider ethics. It is also noteworthy that the Doctor apparently feels no qualms about the human ownership of horses, or the practice of exploiting them for their labor power: all the post 2005 Doctors have all been depicted riding horses and/or riding in horse-drawn carriages.

Returning to the theme of anthropocentrism, the earlier discussion of the Doctor's acceptance of 'the other' also extends to human beings ourselves.

As the 10th Doctor puts it, "Human beings; you're amazing" (*The Impossible Planet* 2006). Despite outward appearances, the Doctor remains an alien Timelord, but his love for his human other in the new series of *Doctor Who* encompasses flirtation, kissing and even romantic love with his human companions. The Doctor's manifest favoritism for human beings over and above other species in *Doctor Who* may make narrative sense given his human appearance and given the devotion which his attractive female companions offer him. However that favoritism spills over into a monolithic speciesism in the Doctor's disregard for the well-being or even survival of nonhuman earthlings. Threats to the survival of planet Earth horrify the Doctor if and only if that destruction also threatens human beings; the planet and its other inhabitants apparently do not concern the Doctor (even though the Earth's destruction would risk cutting off his supply of steak, hotdogs, porky choppies and fish fingers). This is manifested by the 9th Doctor in his second episode, *The End of the World* (2005), set in the far future at the point of the Earth's destruction by an expanding sun, witnessed as a spectacle by alien tourists from space. Overlooking the Earth, Rose asks the Doctor, "what about the people?" to which he replies, grinning, "It's empty. They've all gone. No one left." The 10th Doctor evidences the same comprehensive anthropocentrism in his fight against the Cybermen in *The Age of Steel* (2006). Here he opposes the Cyberleader's vision for a Cyberman-controlled parallel Earth by decrying its lack of, "the *one thing* that makes this planet so alive; people!" (emphasis added). The 11th Doctor repeats the pattern when he refers to the 'homo reptilia' species as having 'occupied the planet before humans' (*Cold Blood* 2010), affirming that no other species of earthlings count or have ever counted as 'occupants' of planet Earth. As in the 3rd Doctor's original encounter with homo reptilia (then referred to as Silurians (*Doctor Who and the Silurians* 1970), the Doctor attempts to broker a peaceful coexistence between them and the human species, again ignoring whatever impacts that might have on other animals. Ironically, this attempt at fostering a peaceful, albeit segregationist, future occurs after the 10th Doctor's rejection of a similar situation: an 'evolved' Dalek plan for sharing the Earth in *Evolution of the Daleks* (2007). In this episode, the Doctor asserts that, "there's no room on Earth for another race of people". Commenting on his own averting of disaster in the episode *Voyage of the Damned* (2007), the 10th Doctor refers to the Earth population as 6 billion, affirming that only human lives enter into his calculations, a point that recurs in the historical episode *The Fires of Pompeii* (2008), when the Doctor is only concerned for the imminent loss of 20,000 human lives in the volcanic eruption. This anthropocentrism (tinged with sexism) is perhaps most starkly expressed in the 9th Doctor episode *Father's Day* (2005), in which the Doctor and Rose travel back in time and witness the moment of Rose's father's death, knocked down by a car. Rose intervenes to prevent her father from being killed. Although this changing of history has dire consequences in the episode, the Doctor comforts Rose by declaring, "Rose, there's a man alive in the world who wasn't alive before, an ordinary man; that's the most important thing in creation". These are examples of the Doctor's repeated insistence on the special status of human beings, which reinscribes the ethical distance between humans and other animals in the programme and facilitates the Doctor's selective exercise of compassion, and his denial of empathy for those animals exploited to provide him, and his audience, with 'meat'.

Conclusion: Challenging 'Meat' Culture in *Doctor Who*

What distinguishes *Doctor Who* as an artifact of popular 'meat' culture is that it transcends the usual fictional treatment of the moral leadership of an eponymous hero; it is a specifically interspecies moral leadership, where the central protagonist is himself nonhuman interacting chiefly with 20th and 21st century humans. Its inconsistencies and contradictions are therefore all the more problematic, because it *invites* consideration of the nonhuman 'other'. However, it overwhelmingly only invites consideration of the nonhumans it *invents*, thereby leaving the speciesist social context of the audience untroubled with regard to their relationship to actual nonhuman species and especially to those they may habitually consume as 'meat'. Matt Smith's tenure as the Doctor was even accompanied by the BBC promoting an annual 'fish fingers and custard day' (BBC 2013a), which elicited the submission of fans' video clips of their creative uses of these foodstuffs, thereby directly connecting the fictive and nonfictive reproduction of 'meat' culture.

It need not be this way. In 1985, vegetarian[9] writer Robert Holmes scripted the story, *The Two Doctors*, starring Colin Baker and Patrick Troughton as the 6th and 2nd Doctors respectively. In the story, the Doctor and his human companions were menaced by alien, but humanoid, Androgums (an anagram of Gourmands). One Androgum character, Shockeye, was characterized as a butcher with an insatiable appetite for flesh. Shockeye himself discovers a cookbook in the kitchen of a Spanish villa (the principle setting for the story)

---

9 Holmes' vegetarianism is confirmed in the DVD production subtitles for *The Two Doctors*, where Nicola Bryant, who played the 6th Doctor's companion Peri in the story, is quoted thus: "As a non-meat eater, I liked the idea. Robert Holmes was a vegetarian, and we talked about it, and that's why he had written the story." (Bryant 2003).

and Holmes uses this as an opportunity to highlight the connection between Shockeye's murderous appetites and the human use of other animals for 'meat', when Shockeye admiringly comments that, "many beasts are bred especially for table" and that, "they [animals] are force fed to improve the flesh, and penned in small confined quarters, to fatten more rapidly". Shockeye later attempts to slaughter companions Peri and Jamie. When Jamie is strapped to Shockeye's makeshift slaughter table, Shockeye discusses "tenderizing the meat" and asserts that this "works better on a live animal", justifying this with the claim that "primitive creatures don't feel pain in the way we would". In the accompanying DVD commentary track, Colin Baker remarks that, "that really is Holmes' message, and it does, er chime doesn't it? 'Cos that's what we habitually say about animals; 'oh they don't feel pain'" (Baker, 2003). At the end of the story, revolted by what he has witnessed, the 6th Doctor declares to Peri, "from now on it's a healthy vegetarian diet, for both of us."

The Doctor's on screen vegetarianism lasted throughout the remainder of the 'classic' series (excepting the 6th Doctor's consumption of 'marsh minnows' while under an alien influence in the 1986 story *Mindwarp*), before being deliberately abandoned by Russell T. Davies, as detailed above. However, the capacity of the Doctor to regenerate, and to develop ethically in response to circumstances, remain core components of the show, and his future veganism remains open to the skill of the programme's script writers. Even in the vegan wilderness of the post-2005 show, examples abound where critique of nonfictional animal exploitation could easily be developed, some of which have been explored above. In the 11th Doctor episode *Amy's Choice* (2010), the Doctor is lampooned by the hostile Dreamlord in a scene in a butcher's shop: "Oh, but you're probably a vegetarian aren't you, you flop-haired wuss". At first glance, this seems evidence of an anti-vegetarian stereotype, examples of which abound in popular culture (see Rosewarne 2013; Cole and Morgan 2011). However, at the end of the story, it emerges that the Dreamlord was an aspect of the Doctor's own persona, facilitated by a mind parasite that "feeds on everything dark in you, gives it a voice, turns it against you". In this light, the Dreamlord's anti-vegetarian butcher persona is established as a 'dark' aspect of the Doctor, implying that vegetarianism is the better part of him. Unfortunately this interpretation is not developed in subsequent episodes, but the potential for the 11th Doctor to learn from the experience in relation to the 'darkness' of 'meat' culture was there. After just four episodes of his first series at the time of writing, the 12th Doctor's eating practices are not yet established (although disappointingly he sips a milky coffee in his fourth episode, *Listen* (2014)), but his first episode included just such an example of the subtle critique of 'meat' culture: In a restaurant scene with his companion Clara, the

Doctor gently chides her hypocrisy when she expresses revulsion at the idea of consuming internal organs, when he says, "you weren't a vegetarian the last time I looked". Capaldi's Doctor has already asked Clara, and by extension the audience, whether he is "a good man" (*Into the Dalek* 2014); we know from Robert Holmes that he has that potential, to extend his ethics of compassion and fully embrace the logic of his love for 'the other', by renouncing 'meat' culture and embracing a vegan future.

References

42. First broadcast 19 May 2007 by the BBC. Directed by Graeme Harper and written by Chris Chibnall.
A Christmas Carol. First broadcast 25 December 2010 by the BBC. Directed by Toby Haynes and written by Steven Moffat.
A Town Called Mercy. First broadcast 15 September 2012 by the BBC. Directed by Saul Metzstein and written by Toby Whithouse.
Adams, Carol J. *The Sexual Politics of Meat: A Feminist-Vegetarian Critical Theory* (20th Anniversary Edition), New York: Continuum, 2004.
Aliens of London. First broadcast 16 April 2005 by the BBC. Directed by Keith Boak and written by Russell T. Davies.
Amy's Choice. First broadcast 15 May 2010 by the BBC. Directed by Catherine Morshead and written by Simon Nye.
An Unearthly Child. First broadcast 23 November–14 December 1963 by the BBC. Directed by Waris Hussein and written by Anthony Coburn and C. E. Webber.
Baker, Colin. "Commentary." Disc 1. *The Two Doctors*. DVD. London: BBC Worldwide Ltd, 2003.
BBC 2013a. *Fish Fingers and Custard Day*. Accessed 30 March 2013. http://www.doctorwho.tv/whats-new/article/fish-fingers-and-custard-day-2013.
BBC 2013b. *Newspaper Review: Horsemeat Scandal Fallout Goes On*. February 17, 2013. Accessed 16 December, 2013. http://www.bbc.co.uk/news/uk-21489560
BBC 2013c. *PMQs: Cameron on Horsemeat Prosecutions and Meat Checks*. February 13, 2013. Accessed 16 December, 2013. http://www.bbc.co.uk/news/uk-politics-21444662.
Blink. First broadcast 9 June 2007 by the BBC. Directed by Hettie MacDonald and written by Steven Moffat.
Boom Town. First broadcast 4 June 2005 by the BBC. Directed by Joe Ahearne and written by Russell T. Davies.
Bryant, Nicola. "Commentary." Disc 1. *The Two Doctors*. DVD. London: BBC Worldwide Ltd, 2003.

Closing Time. First broadcast 24 November 2011 by the BBC. Directed by Steve Hughes and written by Gareth Roberts.

Cold Blood. First broadcast 29 May 2010 by the BBC. Directed by Ashley Way and written by Chris Chibnall.

Cole, Matthew, and Morgan, Karen. "Vegaphobia: Derogatory Discourses of Veganism and the Reproduction of Speciesism in UK National Newspapers." *British Journal of Sociology* 61.1 (2011): 134–53.

Cole, Matthew, and Stewart, Kate. *Our Children and Other Animals: The Cultural Construction of Human-Animal Interaction in Childhood.* Farnham: Ashgate, 2014.

Cornell, Paul. *Human Nature.* London: Virgin Books, 1995.

Davies, Russell. *Doctor Who: Toby Hadoke's Who's Round 54 (April #02).* Accessed 30 May 2014. http://www.bigfinish.com/podcasts/v/doctor-who-toby-hadoke-s-who-s-round-54-april-02

Deep Breath. First broadcast 23 August 2014 by the BBC. Directed by Ben Wheatley and written by Steven Moffat.

Dinosaurs on a Spaceship. First broadcast 8 September 2012 by the BBC. Directed by Saul Metzstein and written by Chris Chibnall.

Doctor Who and the Silurians. First broadcast 31 January–14 March 1970 by the BBC. Directed by Timothy Combe and written by Malcolm Hulke.

Eccleston, Christopher. "Interview with BBC Breakfast". Disc 1. *Doctor Who. The Complete First Series.* London: BBC Worldwide, 2006.

Evolution of the Daleks. First broadcast 28 April 2007 by the BBC. Directed by James Strong and written by Helen Raynor.

Father's Day. First broadcast 14 May 2005 by the BBC. Directed by Joe Ahearne and written by Paul Cornell.

Fear Her. First broadcast 24 June 2006 by the BBC. Directed by Euros Lyn and written by Matthew Graham.

Gardner, Julie. "Dear DVD Viewer..." *Series 1–4 Episode Guide.* Series 1–4 DVD Box Set. London: BBC Worldwide, 2009.

Hide. First broadcast 20 April 2013 by the BBC. Directed by Jamie Payne and written by Neil Cross.

Into the Dalek. First broadcast 30 August 2014 by the BBC. Directed by Ben Wheatley and written by Phil Ford and Steven Moffat.

Let's Kill Hitler. First broadcast 27 August 2011 by the BBC. Directed by Richard Senior and written by Steven Moffat.

Listen. First broadcast 13 September 2014 by the BBC. Directed by Douglas Mackinnon and written by Steven Moffat.

Love and Monsters. First broadcast 17 June 2006 by the BBC. Directed by Dan Zeff and written by Russell T. Davies.

Midnight. First broadcast 14 June 2008 by the BBC. Directed by Alice Troughton and written by Russell T. Davies.

Mindwarp. First broadcast 4 25 October 1986 by the BBC. Directed by Ron Jones and written by Philip Martin.

Morgan, Karen, and Cole, Matthew. "The Discursive Representation of Nonhuman Animals in a Culture of Denial." In *Humans and Other Animals: Critical Perspectives*, edited by R. Carter and N. Charles, 112–132. London: Palgrave, 2011.

New Earth. First broadcast 15 April 2006 by the BBC. Directed by James Hawes and written by Russell T. Davies.

Night Terrors. First broadcast 3 September 2011 by the BBC. Directed by Richard Clark and written by Mark Gatiss.

Nightmare in Silver. First broadcast 11 May 2013 by the BBC. Directed by Stephen Woolfenden and written by Neil Gaiman.

Orthia, Lindy. (ed) *Doctor Who and Race*, Bristol: Intellect, 2013.

Peggs, Kay. *Animals and Sociology*, Basingstoke: Palgrave Macmillan, 2012.

Planet of the Dead. First broadcast 11 April 2009 by the BBC. Directed by James Strong and written by Russell T. Davies and Gareth Roberts.

Planet of the Ood. First broadcast 19 April 2008 by the BBC. Directed by Graham Harper and written by Keith Temple.

Rose. First broadcast 26 March 2005 by the BBC. Directed by Keith Boak and written by Russell T. Davies.

Rosewarne, Lauren. *American Taboo: The Forbidden Words, Unspoken Rules, and Secret Morality of Popular Culture*. Santa Barbara: Praeger, 2013.

Silence in the Library. First broadcast 31 May 2008 by the BBC. Directed by Euros Lyn and written by Steven Moffat.

Stewart, Kate, and Cole, Matthew. "The Conceptual Separation of Food and Animals in Childhood." *Food, Culture and Society* 12.4 (2009): 457–76.

The Age of Steel. First broadcast 20 May 2006 by the BBC. Directed by Graeme Harper and written by Tom MacRae.

The Almost People. First broadcast 28 May 2011 by the BBC. Directed by Julian Simpson and written by Matthew Graham.

The Christmas Invasion. First broadcast 25 December 2005 by the BBC. Directed by James Hawes and written by Russell T. Davies.

The Doctor's Wife. First broadcast 14 May 2011 by the BBC. Directed by Richard Clark and written by Neil Gaiman.

The Eleventh Hour. First broadcast 3 April 2010 by the BBC. Directed by Adam Smith and written by Steven Moffat.

The Empty Child. First broadcast 21 May 2005 by the BBC. Directed by James Hawes and written by Steven Moffat.

The End of the World. First broadcast 2 April 2005 by the BBC. Directed by Euros Lyn and written by Russell T. Davies.

The Fires of Pompeii. First broadcast by 12 April 2008 the BBC. Directed by Colin Teague and written by James Moran.

The God Complex. First broadcast 17 November 2011 by the BBC. Directed by Nick Hurran and written by Toby Whithouse.

The Hungry Earth. First broadcast 22 May 2010 by the BBC. Directed by Ashley Way and written by Chris Chibnall.

The Impossible Planet. First broadcast 3 June 2006 by the BBC. Directed by James Strong and written by Matt Jones.

The Lodger. First broadcast 12 June 2010 by the BBC. Directed by Catherine Morshead and written by Gareth Roberts.

The Parting of the Ways. First broadcast 18 June 2005 by the BBC. Directed by Joe Ahearne and written by Russell T. Davies.

The Power of Three. First broadcast 22 September 2012 by the BBC. Directed by Douglas Mackinnon and written by Chris Chibnall.

The Rebel Flesh. First broadcast 21 May 2011 by the BBC. Directed by Julian Simpson and written by Matthew Graham.

The Runaway Bride. First broadcast 25 December 2006 by the BBC. Directed by Euros Lyn and written by Russell T. Davies.

The Sontaran Experiment. First broadcast 22 February–1 March 1975 by the BBC. Directed by Rodney Bennett and written by Bob Baker and Dave Martin.

The Time of the Doctor. First broadcast 25 December 2013 by the BBC. Directed by Jamie Payne and written by Steven Moffat.

The Time Warrior. First broadcast 15 December 1973–5 January 1974 by the BBC. Directed by Alan Bromley and written by Robert Holmes.

The Two Doctors. First broadcast 16 February–2 March 1985 by the BBC. Directed by Peter Moffatt and written by Robert Holmes.

The Unicorn and the Wasp. First broadcast 17 May 2008 by the BBC. Directed by Graeme Harper and written by Gareth Roberts.

The Unquiet Dead. First broadcast 9 April 2005 by the BBC. Directed by Euros Lyn and written by Mark Gatiss.

The Vegan Option 2013. *Science Fiction and Animals*. Accessed 9 September 2014. http://theveganoption.org/2013/07/05/science-fiction-doctor-who-hg-wells-swift-frankenstein-under-skin-planet-apes-trek/

The Waters of Mars. First broadcast 15 November 2009 by the BBC. Directed by Graeme Harper and written by Russell T. Davies and Phil Ford.

Tooth and Claw. First broadcast 22 April 2006 by the BBC. Directed by Euros Lyn and written by Russell T. Davies.

Voyage of the Damned. First broadcast 25 December 2007 by the BBC. Directed by James Strong and written by Russell T. Davies.

Wells, H. G. *The War of the Worlds* (New Ed edition), Penguin Classics, 2005.

World War Three. First broadcast 23 April 2005 by the BBC. Directed by Keith Boak and written by Russell T. Davies.

CHAPTER 12

On Ambivalence and Resistance: Carnism and Diet in Multi-species Households

Erika Cudworth

Introduction

Particularly in richer parts of the globe significant numbers of us keep a small number of certain species as companions in our homes. Sixty three per cent of North Americans live with a companion animal, and in 2006, spent $40.8 billion on food, bedding, toys, and recreational activities with their animals (Williams and DeMello 2007, 231–32). In New Zealand, the figure is even higher with 68 per cent of households including a companion animal—the highest level of companion animal guardianship in the world (New Zealand Companion Animal Council 2011, 8).

In the United Kingdom (UK), almost one in two households include a companion animal (46 per cent, Pet Food Manufacturers Association, 2014) with the most popular companions being dogs (9 million) and cats (8 million) (Pet Food Manufacturers Association, 2014), which together account for around 45 per cent of animals kept as 'pets' (Pet Health Council 2008). Overwhelmingly, these non-human animals depend on 'their' humans for food.

Until the late nineteenth century, domestic dogs were fed leftover food from human meals and stale or rotten food, and this is still common practice in 'developing' countries. From the development of dog 'biscuits' in the 1860s, a global industry has emerged which utilizes enormous quantities of readily available corn, wheat, rice, potatoes and soy as well as huge quantities of blood, bone, skin and flesh from domesticate animals slaughtered for meat. In 2007, the pet food industry in the United States (USA) alone sold over $16 billion worth of 'goods' (Williams and DeMello 2007, 231). While there are various manufacturers of pet food, the industry is dominated by major food companies and pet food has become a lucrative market for extracting profit by recycling gluts of grain and animal parts deemed unsuitable for human consumption. Pet food is an important element of the global political economy of meat. In this context, this chapter explores the role of carnism in the feeding of companion animals.

Pet food is the subject of vigorous advertising encouraging human consumers to buy products that constitute a healthy, natural and/or economical diet for their companion cat or dog, and thereby demonstrate care and responsibility. In such advertising, animal 'actors' make their preferences clear. In the current Iams cat food advertising campaign on British television, the feline actor is blunt: "I am not a vegetarian", says the voice-over, "these teeth were made for meat" (Iams 2012). The consumption of this food is thereby naturalized for animal companions. Research on the relationship between 'pet ownership' and concern with the treatment of non-human animals more widely, is inconclusive, although there is some evidence that living with a companion animal may encourage empathy to other kinds of animal (Serpell and Paul 1994). This chapter draws on an empirical study of human relations with companion dogs in which many human participants did demonstrate concern beyond 'their' animal, with the treatment of 'pet' animals (dog fighting, the use of dogs in baiting and hunting wild animals, animal breeding) and animals in agriculture (particularly intensive farming). Although the majority of interviewees in this study were meat-eaters, many did place some restriction on what they ate in terms of the creatures they would eat, or the methods of production involved in the animal foods they ate. Whilst many were troubled by the 'what we love/ what we eat' distinction, even the vegetarian dog 'owners'[1] and the lone vegan in the study all fed their dogs meat. Carnism was operationalized unproblematically in discussing dogs' food and diet. Narratives of human choice infuse practices of dietary resistance to carnism, whereas biologism effectively reproduces the 'love/eat' distinction when it comes to feeding animal companions. As one interviewee put it, 'I'm a vegetarian, but she's a dog'.

This chapter proceeds through five sections. The first two sections set the context. First, a short mapping of the development of pet food, ending with a product recall which threw relations between humans, pets and those few species that are routinely eaten, into sharp relief. Second, I examine the normativity of consuming a limited range of animals in contemporary (Western) cultures through the concept and practices of 'carnism'. This, Melanie Joy

---

1 Interestingly, whether vegan, vegetarian or carnist, all interviewees referred to themselves and others as dog owners. None used the term 'guardian'. While in Critical Animal Studies, the term 'guardian' is often preferred, I have come to wonder about this. In countries such as the UK where animal companions are legally property, the term 'owner' is both an accurate description and perhaps a continuous reminder of the vulnerability of animal companion under the law and dependent on the attitudes and behavior of their 'owners'.

(2010, 28–9) suggests, is so pervasive in our culture that we see it as routine, normative, apolitical—'meat eating', whereas eating plants becomes a political choice for 'vegetarians' and, more deviantly, 'vegans'. The second part of the paper uses material from an ethnographic project involving people who live with dogs. It will be argued that carnism pervades discourses of care and concern for much loved dogs and that there are often also ambiguous relations to the animals such dogs eat. It will be argued that even for vegetarian and vegan 'dog owners' carnist discourses reinscribe meat eating practices for the non-human members of the multi-species household.

Dominance and Affection: The Making of Pet Food

For Yi-Fu Tuan (1984), it is a complex relationship between domination and affection that leads to the historical development of different categories of 'pet'. It might also be said however, that similar forces are key to the development of 'pet food'. There is often seen to be a fundamental ambiguity at the heart of our treatment of the small range of species kept as domesticates. On the one hand, animals are exploited in their billions for food, and on the other, a small minority of domesticates are kept as companions in homes where they may be unlucky enough to be mistreated or lucky enough to have owners who care for them greatly (see Cudworth 2011, 5–8).

There are ways in which the keeping of animals as companions both replicates in minor form, the exploitative breeding conditions of the meat industry (Williams and DeMello 2007, 236–47). In addition, 'livestock' farming and pet keeping are integrated forms of human domination of domesticates, for the breeding and selling of animals as pets ultimately supports a multi-billion pet food industry which is an important and highly profitable part of twenty first century agri-business (Nibert 2002, 92–95). Certainly, every British slaughterhouse has a 'pet food bin' for the meat unfit for the McDonalds skip[2] or for regular human consumption (Cudworth 2008). The amount of 'resources' and the potential profit to be obtained from them is considerable:

---

2 This term is used within UK slaughterhouses to describe the skip where the workers throw the parts of animals that will be turned into processed meat products such as burgers and sausages.

> Whole mammal bodies (cows, sheep, pigs etc.) by the hundreds of thousands, millions of their major parts and thousands of tons of bird flesh end up in the reject pile. Much of this waste is diseased, and often cancerous... This reject pile is soon on the move as it magically changes into "meat meal" and "by-products" on pet food labels (Peden quoted in Nibert 2002, 94).

Pet foods have always been produced from the leftover parts of animals which are not to be used for human food such as bones, organs, ears, skin and so on. However, the industry is much expanded since the mid twentieth century. In the 1960s, most 'pet' cats and dogs were fed dry biscuits, table scraps and some standardized forms of canned food produced from slaughterhouse by-products (Franklin 1999, 90). By the 1980s, pet food manufacturers were realizing the potential of diversification and offering an array of brands, all produced by a small number of firms, to appeal for a range of human (rather than non-human animal) consumers (see Nestle 2008).

In 2007, the pet food industry in the US alone had sales of $16.1 billion (Nestle 2008, 42). While there are many companies which manufacture pet foods, the industry is dominated by five of these: Nestlé (with Purina Pet Care brands), Mars, Proctor and Gamble (Iams), Colgate (Hills 'Science Diet') and Del Monte (Nestle 2008, 43). Most pet foods have complicated lists of ingredients including, along with meat, fish and a range of animal derived products, thickeners and stabilizers, vegetables and cereals and a host of other additives to improve color, flavor and texture. Commercial pet foods are tested on animals,[3] often unwanted and surrendered to dog pounds, or stray animals. The ingredients of pet foods, their journey 'from farm to bowl' and the animal testing involved with most large corporate enterprises, are obscured by the packaging and advertising of pet foods. Rather, advertising campaigns specifically target positive relations of care and concern that human 'owners' have for the cats and dogs they assume responsibility for feeding. In the United States, a Proctor and Gamble campaign for its premium Iams range brought this to the fore:

> The unconditional love between people and their pets is at the center of the new Iams marketing campaign—Keep Love Strong. The new advertising campaign, created by Saatchi & Saatchi out of New York, will

---

3 Various kinds of tests might be used, from the more benign taste trails, to toxicity testing whereby dogs will be made ill, often critically so, and in some forms of testing (such as 'lethal dose') will die (Williams and DeMello 2007, 198).

feature both television and print ads showcasing the important role premium nutrition like Iams plays in keeping a dog or cat's body as strong as their love.

"Every day we're inspired by the stories we hear from our consumers about the love they have for their pets and the important role Iams plays in those relationships," said Ondrea Francy, Iams general manager. "One of the most exciting things about our new campaign is that it was all inspired by real stories of unconditional love." (Proctor and Gamble press release 2012)

Chad Lavin (2013) introduces the concept of 'digestive subjectivity' in order to think about the politics of consumerism and public health, understandings of political identity, the operation of political power and the processes and ideologies associated with modernity, democracy and globalization. Contemporary discourses promoting pet foods can be understood in Lavin's terms of an emphasis on individual responsibility and a shifting of political terrain towards the fashioning of "entrepreneurial subjects" who engage in self-surveillance. In the case of pet foods, the 'owner' enacts surveillance in terms of their own identity (as a 'good', loving or responsible pet owner) and in terms of selecting appropriately from an ever-expanding range of products. Food is highly political, and contemporary food discourse is fraught with different kinds of anxieties—around health, weight, ethics or identity, in Lavin's account. These anxieties are bound up with those around globalization wherein many people feel that their lives (and the foods they consume) are beyond their control. As such, we have become, certainly in the West, 'digestive subjects' fraught with anxieties about the embodied vulnerability that attends our consumption of food. 'Food' as Lavin notes, "is rarely only food" but is political in a number of ways. At one level, a range of anxieties are associated with the food we eat—we are threatened by the presence of pesticides, trans-fats, salt and super-sized portions. As a result, consumers require protection from corporations and also from themselves via political intervention. Also political is the production of food within networks of globalized agribusiness that undermines the regionalism of diet, the link between diet and national identity and, potentially, our safety in long supply chains with limited accountability. In the case of pet foods, owners are responsible for the health and welfare of the animals in their care, and care is practically and emotionally demonstrated, by feeding. But what happens when Iams does not keep cat and dog bodies 'as strong' as their owners' 'love'?

Anxieties in relation to the 'politics of petfood' have been considerably heightened since the 'pet food recall' scandal of 2007 in the US. For two months, from March 2007, a significant number of leading dog and cat food manufacturers recalled products from retailers. The products recalled included familiar premium brands like Eukanuba, Iams, Hill's Prescription and Royal Canin. This was the largest recall of consumer products in US history. In the aftermath, large numbers of canine and feline animal companions (4–5, 000) were found to have died from severe kidney failure resultant from a toxic chemical cocktail present in their food. US pet owners' faith in many pet food manufacturers had been shaken, and contamination had been traced back through the supply chain revealing connections in the sources for pet food and products fed to agricultural animals including the huge numbers of animals 'grown' for meat, milk and eggs to be consumed by human beings. In the US, pet food sales fell by 30%, the Food and Drug Administration was in crisis and lawyers and animal advocates were attempting to "take pet liability damages into the stratospheric heights of compensation for human liability" (Nestle 2008, 162).

The food industry routinely uses cost cutting initiatives, and adding waste products from one kind of industrial process such as melamine manufacture, to boost nitrogen levels in animal feed—including pet food, baby milk and children's food products for example, has been common. However, it was the effect of combining melamine with cyranic acid, another nitrogen rich cheap chemical, which produced crystals in animals' urine and precipitated rapid and fatal kidney damage (Nestle 2008, 27–52). The outbreak was caused by adulterated wheat gluten and rice-protein concentrate from China (see 2008, 88–92). The strength of feeling that the 'pet community' in the US have for their pets is not to be underestimated—pet owners jammed veterinary surgeries and phone lines, set up internet sites and ultimately a 'Good Pet Food' movement. At the close of this recall, the company at fault, Menu Foods, agreed to compensate owners of dead pets for their losses, in terms of the financial costs they incurred (of veterinary bills, cost of purchase and replacement of the animal and so on) (Torres 2007, 59). A total of $12.4 million was re-imbursed, about half of that which was claimed (Lau 2011). While legally in the United States, animals are property, and might simply be replaced like a damaged piece of household furniture, that is clearly not how they were viewed by those who lost the dogs and cats of their hearts to kidney failure, and in some (few) jurisdictions, such as Washington DC, successful claims were made including compensation for 'emotional distress'.

The pet food recall demonstrated the global scope of the complicated networks of the food industry in which 'waste' products from the human food chain and other productive processes are recycled. One response to the recall was an increase in vegetarian pet food, another was a more powerful voice for the 'raw food' and 'natural diet' groups often lobbying consumers via websites on healthy feeding of pets. There are certainly alternatives to meat-based diets for animal companions. In the UK, vegan veterinarian Andrew Knight has been a pioneering force, drawing together research on the health benefits of meat free diets for dogs and cats (respectively, Brown et al., 2009; Wakefield et al., 2006) and defending veg(etari)an diets for companion animals as at least as healthy, if not healthier (Knight, 2005, 2008). The website 'vegepets' promotes such diets, recommending suppliers, brands and diets for animal companions (see www.vegepets.com). However, this is a limited commercial presence in the UK. Iams, along with other products involved in the pet food recall, have recovered and often expanded their markets, demonstrating also that although faith in the production of animal-based foods for companion animals was undermined in 2007, in the long term, belief in the necessity of animal-based foods for companion animals was not. The next section considers an explanation for why that might be.

Carnism and Animal Companions

As Melanie Joy points out, all cultures demarcate species that cannot be eaten and those which can (2010, 13). These differ across cultures so that while in India both dogs and cattle live either domesticate or feral existences, it is cattle in Hindu dominated states that are protected and may not be eaten. In the West however, it is common practice to feed cattle-meat to dogs kept as companions in the home. A common Western response to the idea of eating dog is disgust, but what is of most significance here, Joy challenges, is that most people do not experience disgust at the thought of eating the very small range of animals which Western cultures deem edible. We do not experience disgust because meat eating is an established social practice and normative in our culture. We believe it is legitimate—both ethical and appropriate—to eat animals such as cows, pigs and sheep and this constitutes a belief system she calls 'carnism' (2010, 29).

Carnist belief systems are predicated on a series of myths revolving around what Joy terms the three 'N's' of justification: that eating meat is normal, natural and necessary. These ideas are so ingrained in our consciousness that we do not have to think about them—we have internalized them as truths (2010, 95–7). Professionals such as veterinarians, doctors and non-government organizations (ngos) supporting corporate agribusiness represent institutionalized carnism and function as "emissaries for [carnist] myths" whilst the mass media "reinforce the carnist message of the way things are" (2010, 100 and 105). Whilst conformity is culturally rewarded, deviance is punished and thus "vegetarians often find themselves having to defend their choices, explain their diet, and apologise for inconveniencing others. They are called hypocrites if they wear leather, purists or extremists if they don't." (Joy 2010, 106). Joy makes the simple but important point that in a carnist society where meat eating is customary and 'traditional', its longevity as an established cultural practice makes it easier to justify and far more difficult to question. A second key form of justification is 'naturalization'. Practices such as meat eating are seen to be in line with the laws of nature and having a biological basis (2010, 107–8). Third, and closely connected, is the belief that eating meat is necessary, particularly for health, but also for economic growth and stability (2010, 109–13).

Joy considers carnism to be a system whose beliefs we internalize. In the process, the animals eaten as meat are objectified and de-individualized. We also categorize animals into those we are prepared to eat and not prepared to eat based on traditional practices and on ideas about appearance of animals (there are those who will not eat rabbits but will eat turkeys) or intelligence (those who will not eat pigs or dolphins but will eat sheep) or those animals often considered disgusting (rats, snakes, pigeons) (2010, 114–28). What is needed, in Joy's view, is to (re)awaken empathy for animals and invigorate disgust for the meat offered on a plate. However, Joy underestimates the ability of systemic social relations to reproduce themselves here. It is not enough to 'crack' the carnist system and assume that once empathy and disgust are engendered, a moral compass in which all animal products are eschewed follows. Rather, the questions of boundary drawing are more complicated. Erica Fudge (2008, 18–19) emphasizes this reinscribing of boundaries in terms of what species of animal are and are not eaten. For vegans, boundaries are clearest of all, and for vegetarians they are reasonably clear in that "beings with animate life" are "not to be eaten". It is meat eaters who live, she says, with boundary confusion, continually making decisions about "where the boundary between the edible and inedible exists". Both Joy and Fudge underestimate however, how the boundaries are blurred even for some vegetarians and vegans. Those with a plant-based diet inevitably consume animals—they may accidentally and routinely eat non-plant micro-organisms. I do not consider this ethically problematic as the question for ethical vegans focuses not on the avoidance of the consumption of the living per se, but on avoiding forms of exploitation and intentional harm towards nonhuman animals and promoting more compassionate ways of living.[4] However, for those who live with companion animals, the questions of 'who' eats 'what' mammals, birds and fish, are problematic, particularly for the majority of companion animal 'owners' who live with dogs (biologically classified as carnivores, but often seen as omnivores with a physical and undeniable carnivorous bias) and cats (classified as carnivores and often seen as 'obligate' carnivores).

The remainder of this chapter considers how carnist ideas might be seen to underpin our processes of categorizing non-human animals. It looks at the ways in which these categorizations might be shifted and disturbed by the practice of living with an animal companion, and also considers how such categories find ways of reconfiguring. The data is drawn not from studies I have undertaken on the production and consumption of meat (Cudworth 2008, 2010), but from my latest work looking at relationships between people and dogs. In interviews focusing on their lives with dogs, many people discussed the food they fed their dogs and often this also spilled over into talk of what they ate themselves. The majority of interviewees were 'carnists', but many of these placed restrictions on the range of animals they would eat and many were also concerned for the welfare of animals other than dogs. In this sense, they are dealing with the problematic boundaries of the animals we may and may not eat, and for many, this was troubling. Most interesting were the vegetarian and (lone) vegan dog owners for whom one might imagine the carnist glass to have been cracked. However, in every case, companion animals in the home were fed meat. The empirical material suggests that carnism reinscribes itself in the ways in which veg(etari)ans living with dogs discuss the ambiguities, tensions and reconciliations between their belief systems and the practice of feeding meat to the dogs-of-their-heart.

---

4   This is debatable and there are those who would argue that in avoiding harm to the smallest of creatures we might wear face masks to avoid ingesting them, or sweep our paths so that we do not walk on them. This has been associated with Jainism, of course. More secular perspectives in the West might suggest that to avoid creating 'road kill', veg(eteri)ans should not drive cars. I think however there is a distinction between intentional and non-intentional killing. For example, whilst ethical vegans would avoid supporting intentional killing (the use of insect bodies in producing cosmetics or food dyes, or consuming insects as food items), accidental killing of tiny creatures, often invisible to us, is unavoidable.

A Note on Methods

The material that follows is taken from two ethnographic sources—interviews with people who live with dogs, and observations which were recorded. Field notes were kept in the form of an ethnographic diary of encounters with dogs and their people in part of London's Lea Valley Park, recorded daily for a calendar year across 2009–10. In addition, material is drawn from semi-structured interviews with dog guardians, investigating their relationships and everyday lives with canine companions. Thirty seven interviews were undertaken in 2010 and 2011 with people walking dogs on the marshes, which form part of the Lea Valley Park. A second phase of interviewing was undertaken in 2014 to see how far locality affected the data, and fifteen interviews were undertaken with people walking dogs in and around a village in rural Leicestershire. There were, surprisingly, no substantial differences in data obtained from interviewees living in urban and rural locations. One difference of note for this chapter however, is that the vegetarians were more strongly represented amongst the London-based interviewees. The overwhelming majority of interviewees were white, a reflection of the dog-walking population in both locations.[5] Most interviewees were women (three quarters of the sample) which again, is an accurate reflection of the human population walking with dogs in both study sites and may perhaps explain why the number of vegetarians in the sample was much higher than might have been expected. The social class background of the human participants is difficult to ascertain (except where interviewees explicitly discussed their background and/or occupation, and/or where in the London context, class might be inferred from accent).[6]

Most of the interviews were of the 'walk and talk' variety, a combination of participant observation with interviewing. The idea is to accompany informants as they go about their daily routines, asking questions about their lives along the way (Hall, Lashua and Coffey 2006). Some using this method are very interested in the effect of space on the nature of what people say and thus tend to have fixed routes (Jones et al. 2008), whereas this project followed participants on their usual route, the priority being to put people at ease and talk to

---

[5] In the rural location, there are only two non-white people walking with dogs. At the time of the study, one did not yet live with dog companions, and I had yet to encounter the other!

[6] This might also be inferred from accents from the midland of England, but this author, being a 'Londoner' and relatively unfamiliar with different East midlands accents, would not be able to differentiate.

them about their dogs in a situation where they are 'dog-focused'. Some interviewees chose other locations, such as pubs, cafes or their own homes.

The interviews chart the practices of 'responsible' dog owners who walk their dogs very regularly, interact with them and have close bonds with them. Those who do not walk with dogs were not part of the communities studied, neither were those referred to by regular dog walkers in both the research locations as the 'fair weather walkers' who walk when they feel like doing so, rather than as a daily practice. This may well influence the findings presented here—participants in this study already demonstrate certain dedicated practices of care for the dogs they live with by regularly exercising them. Various themes emerge in the data and many revolve around practices of care for companion dogs. The 'responsible' dog owners in this study were concerned about the treatment of some other animals. Some were concerned about the well-being of 'wild' animals, some about the keeping of certain animals as pets (caged rabbits, in particular). Many interviewees had relationships with other animals in addition to dogs (birds, cats, rabbits, guinea pigs, horses and goats) and one of the interviewees from the urban sample and three from the rural sample were brought up on farms as children. I had expected that an association with farming may have influenced attitudes to dogs amongst these latter interviewees, and had expected a more utilitarian perspective, but was surprised to find that this was in no way the case. In fact, all these interviewees were in the sixty per cent of the sample who shared their bed with dogs at night.

In the course of these discussions, negative views were often expressed about industrial animal agriculture ('factory farming') and industrial models of pet production ('puppy farming'). I would suggest therefore, that living with dogs involves both ambivalence and a reinscribing of species boundaries (between dogs and farmed animals) *and*, although to a lesser degree, a broadening of concern with issues of animal welfare. This is very much evident when it comes to the discussion of diet. In the quoted material below I have chosen to differentiate between interviewees as those meat-eaters who place some restrictions on what they will eat (restricted carnists) and those who do not (unrestricted carnists) in addition to those who describe themselves as vegetarian or vegan.

I'm a Vegetarian, but She's a Dog

A question often asked of companion animal ownership is whether it enhances or augments a wider concern with animal welfare or even an understanding of animal rights, or whether it places lovers of dogs in a situation of ambivalence as they continue to eat other animals. For some who live with dogs, and for whom eating meat is ethically unproblematic, discussions of what companion dogs eat reveal no sense of anxiety around species boundaries. In this quote below, the anxiety focuses on which food will not spark an allergic reaction rather than 'what' is being eaten:

> Everythink's[7] got to be fish-based, so we've got him on 'Chappie' and it took a lot of time to find one that didn't flare his ears and his eyes up. The vet suggested we put him on 'Hills', which sent him hyperactive—sixty quid a bag, as well. It sent him hyperactive so I whisked him off of that. We also had one called 'Wafcol' but it made his ears quite bad so we experimented, and we went down to [a large pet store]. We get their own brand of the fish and rice, the dried food. So he 'as that and 'Chappie'. He can have pigs' ears for treats, pork strips for treats, anything pork or ham he's fine with. (unrestricted carnist)

However, issues of animal welfare did surface in many of the interviews, in particular around the use of animals as food. Where this was raised as an issue by interviewees, it was in terms of their own vegetarianism or the expression of conflicted feelings around eating animals, and in many cases this was linked to ideas about people's relationships with their companion dogs. In another case of a dog with a restricted diet because of allergies, there is a very different narrative to that of the interviewee above, this time from a vegetarian 'owner':

> It just upsets me so I have not to think about it because it's just awful. We've tried everything and it's hopeless. We've done all the kinds of vegetarian food, we tried lots of different fish, fish and rice-based ones and that all made him really ill. He can't eat beef, he can't eat chicken, it has to be pork or he is ill. And that's just about the worse thing, isn't it? It's just the worst thing and god knows how it's produced. Well, we do know don't we? In factory farms, awful. So [name of dog] eats pigs and I hate it. I hate it. (vegetarian)

Clearly here, there is a hierarchy implied between species in terms of the 'least poor' ethical choices that this owner might make. Despite the tension in the

---

7 Please note that the East London (cockney) accent has been retained in quotes where applicable.

narrative, any dilemma is resolved in favor of the health of the dog rather than "make him ill". Important to note is that this interviewee considered the least ethical option would be to give the dog up for rehoming so that "someone else gets to buy all that stuff and not me". In another case, a similar hierarchy of 'least poor' ethical choices is both resisted and legitimated by recourse to a rather odd form of naturalization:

> I don't think it is more acceptable for a chicken to be slaughtered than a cow, but somehow I can, I can handle the thought of [name of dog] as a dog, eating a bird more than I can handle the thought of him eating a lamb or a cow because he's so small I think. I don't know, but there's something there, I don't know what the logic is but there is, there's something that always sends me to the chicken-based thing [dog food]. (vegetarian)

It is not only vegetarians for whom the boundaries between different animals and what may/not be eaten become problematic. In this discussion between two interviewees (a couple) the questions of species and boundaries are brought into sharp relief when they move from talking about what their dogs eat, to what they eat, with disgust, naturalization, denial and anxiety all making their presence felt:

> J: well I do eat other meat [than chicken-meat] but I do find it hard not to be sick sometimes when I think of what you're eating. I haven't got any conscience where chicken's concerned. It's horrible really.

> S: I'd be more concerned about eating chicken, because of the welfare of the chickens rather than cows. I mean, I'm not an expert, but I imagine them to be romping round the fields, before, you know, at least for a bit you know, more than your average factory chicken... I mean it's horrible isn't it? I mean, we're not vegetarian but we don't eat much meat that's not white meat. Although, you know, I don't think it's necessarily wrong to eat meat and um I've had rabbit and you know other things that may have had a better quality of life, but I suppose I'm like a lot of people, I just push it out of my mind all the time. And pork is probably, well, although it's one of my favourite meats it's probably the one that troubles me the most because they're supposed to be as intelligent as dogs and have similar emotions.... which is quite chilling really. (restricted carnists)

For a number of interviewees, discussions of everyday lives and relations with dogs leads them into a comment on their own dietary practices and often, feelings of guilt are expressed over eating meat:

> I carry it around all the time, the guilt; looking for organic and free range. I don't eat much meat and when I do I feel guilty. I should give it up really, it's bloody ridiculous. (restricted carnist)

Or alternatively, a general love of 'animals' or 'wildlife' raises this tension:

> When I go to the allotment I usually leave something out for the fox, especially if it's winter time. I don't know, I just love being around animals and I do feel guilty actually, that I eat them. It's a bit of a quandary. There's a little robin that comes and gets worms when I'm digging. (restricted carnist)

All the dogs that interviewees lived with ate meat however, and this was both normalized and naturalized with a form of biological essentialism advanced to justify a diet of meat for dogs. A common sentiment expressed by those interviewees who did not eat meat, was "I'm a vegetarian, but s/he's a dog". Also important therefore, is the notion that human beings are different from other animals in their ability to exercise choice over what to eat and what not to eat:

> It's not a conflict, I don't mind getting meat for her, she's a dog so I don't see that as a problem. I don't mind other people eating meat—if they are eating meat it's their choice. (vegetarian)

> ...the thing about vegetarianism is that it's to do with humans having a choice and not being under any necessity of eating meat. Dogs are essentially wolves and it's an essential part of their diet. (vegetarian)

One quarter of the interview sample were vegetarian and a number made a direct connection between their diet and living with a dog:

> But I do think that living with an animal does raise interesting issues about eating. I mean for me, I find it hard to believe that lots of people live with animals but quite happily eat other animals too. To me, being that close to an animal [his dog] sort of reinforces the idea that for us, for humans, we have relationships with other animals or we're capable of

> having relationships with other animals which are different to those of some other animals. I know loads of people who do live with animals either as pets or as farmers, and happily eat them, but certainly with me, it reinforces my natural inclination to stick to grass. (vegetarian)

> I know loads of people who absolutely adore, love animals and they are very bothered about animal welfare but they eat meat. But... you always think that thing of 'I could not eat my animal, they're animals and I couldn't eat my dog', I just couldn't... because you make that connection, that's an animal that you are eating, I think, I mean lots of people just put it to the back of their mind probably, but when you're biting into that meat, I just imagine that it's [name of dog]—yuk. (vegan)

For vegetarians, feeding meat to dogs was discussed through a tension between the carnist assumptions that dogs needed to eat meat, alongside vegetarian repulsion at what (or 'who'?) was being decanted from the can into the bowl:

> Well, I can cope with it, you know, just about. I wasn't brought up as a vegetarian as a child but [name of partner] was, and he can't really cope with meat at all. I know when he's feeding the dog because I can hear him heaving in the kitchen. (vegetarian)

> ...I do give him like dried dog food but I can't cope with the kind of canned sort of meat thing and I think it is better for him to have like proper food rather than out of a tin but when he's with my ex-partner—he has him two days a week—then he gets tinned dog food. (vegetarian)

For those who cooked food for their dogs in addition to or as an alternative to commercial dog food, or who feed dogs a 'raw food' meat-based diet, there were further issues. For vegetarians, this meant buying and cooking meat they would not otherwise buy, and occupying spaces they would want to avoid such as butchers' shops and fishmongers in search of 'cheaper' or better quality meat. Cooking meat raised particular conflicts where normative and natural discourses of carnism are brought into proximity with vegetarian disgust:

> I don't actually buy free range chicken I buy just bog standard chicken. That goes against all my principles, not buying free-range chicken. One thing is buying the meat in the first place and then I'm not even buying free-range chicken. You know, there's only so far you can go... The stuff I

> buy her is chicken breast, 'cos although I could get it cheaper if it was on the bone—I'm prepared to cut up chicken but I do it with a fork and scissors, but I don't touch it with my hands—but there's no way I'm going to spend time pulling or cutting meat off bones. (vegetarian)

> I have to limit the amount of meat I can give them just because I can't pick the chicken off the bone. (vegetarian)

For carnists, the cooking and preparation of meat for dogs presents no such challenges and this is far more readily integrated into daily routines:

> ...every week they have a cooked chicken. They don't eat the bones or anything like that...we hand strip the chicken and they have that, we cook lamb mince as well so you know and obviously if we're eating we'll give them something but it will only be [a small amount], I'll try and put vegetables in their food, put gravy in it just to give them a bit of vitamin and stuff like that so erm you know they have their own type of food... I adapt our human food to suit their diet really. (carnist)

Vegetarian disgust is overcome by a combination of carnist discourse of naturalization, necessity and normalization, combined with an inflection of dogs as active decision makers in the process, 'choosing' to eat meat. In some cases, this is based on the behavior of dogs when presented with a variety of foods:

> I wouldn't ask them to be vegetarian, they really do enjoy their meat, and I, just because I choose to be vegetarian there's no reason why I expect my dogs to be vegetarian and they love their meat, absolutely adore it..., but I would like, ideally, I would like them to have erm organic chicken, organic meat, plus a reasonable amount of vegetables, but they're very naughty on eating vegetables...their choice is not to eat vegetables, be vegetarian and therefore just because I choose to be vegetarian there is no way I would ask my dogs to be vegetarian. That's my choice, I'm not gonna impose my choices on them. (vegetarian)

> ...having thought about it I decided that I didn't think that decision was really mine, mine to make that I should enforce my dog to have a meat free diet...I spoke to my friend who was a vet and he was very against the idea of him being a vegetarian and I decided actually that is quite, might

> be quite a erm controlling thing for me to do. My dog will be vegetarian when he clearly really likes meat, yeah erm yeah we've tried we've given him vegetarian treats ages ago and he'd just go you know he wasn't interested so, so no, he's he's, he will always eat meat. (vegetarian)

Ultimately, carnism and the notion that dogs have a strong preference for eating other animals, overcomes vegetarian dog owners' conflicted and sometimes very troubled feelings. For most, there is the notion that for dogs, eating meat is 'meant to be' despite many acknowledgements of the problems of commercial pet foods (on both health and ethical grounds) and the dilemmas faced by vegetarians in preparing/cooking meat for canine companions:

> There is, there's a contradiction, there's lots of stuff, y'unno I, there's lots of literature about vegan diets for dogs but I'm not, I don't know [pause] I'm not convinced by it at all. I don't know about a vegetarian dog... they're not the same as us, they don't have the same enzymes... I don't even think if they are always vegetarian [from a puppy] it would work... It's like trying to give a horse [some] meat or a sheep a nice bit of steak, I think there some things that are meant to be. 'You like your meat don't you?' [directed at the dog]. (vegan)

> At first I found it quite hard but I suppose I make a compromise and like I don't, I tend to buy organic, 'happy' chicken, they had a happy life, you know, even if it was very short and it makes me feel a little bit better about it. But I kind of accept it, you know, he's a dog and he's not he's not a vegetarian erm and I think the whole you know the raw meat and the bone and stuff is really good, good for him so I think, I just think his health is more important really so, so he gets well fed and never gets fat. (vegetarian)

Certainly all the dog owners I spoke with were concerned with the health of their animal and considered that this must be a priority. In some cases the dissemination of carnism was re-enforced through the advice of veterinarians. Some of the vegetarians interviewed made specific points about pet food corporations from whom they would not buy "because they do horrible things to the animals they test in their labs". Yet for one vegetarian interviewee, the opinion of an expert overrides this critical knowledge: "I buy Iams because our friend that's a vet says that's the very best food you can buy for them because it's already nutritionally balanced".

Coda: "I Don't Know about a Vegetarian Dog"

The responsible dog owners in this study cared very much about the health, welfare and happiness of the dogs with whom they shared their lives. They had considerable amounts of what Leslie Irvine (2004, 66) calls 'animal capital'—resources that enable consciously less exploitative relations with animal companions such as knowledge about training and health or an active interest in animal emotions or cognition. The interviewees also had concerns for animals other than their own dogs, certainly for dogs in general, for animals kept as companions more widely, for wild animals and in some cases, for non-companionate domesticate animals. For the 'unrestricted carnists' issues of an appropriate diet for their dogs was unproblematic, but the majority of people interviewed did restrict their carnism in certain ways and their discussions of their own diets and that of the dogs they live with reveal tensions and uncertainties regarding what animals can and should be eaten as food. For the vegetarians, ambivalence was heightened.

All the vegetarians with whom I spoke claimed to 'love animals' and this, often alongside a history of having close relationships with dogs as companions, was given as a reason for their own choice of diet. All of them bought and fed meat—sometimes that they had prepared/cooked themselves—to the dogs in their care. The 'three N's' of carnism make their presence felt in discussions of animal diet. Feeding dogs meat is seen as essential for their wellbeing (in the case of those with dietary allergies) or more broadly in terms of the most nutritious and healthy diets possible. Despite rejections of carnism in their own diets, these interviewees normalized and naturalized a meat diet for dogs by drawing on arguments for biological difference and physiological necessity. As the lone vegan in the study put it: "some things are just meant to be".

My vegetarian interviewees were not naïve—many had considered a vegetarian or vegan diet for their dogs and some had experimented with such diets before deciding to reject them. Clearly some of them felt the pressures of carnist normalcy and felt that there was no other decision to be made than feeding their dogs a meat-based diet. There are dogs (and cats) living well on vegetarian and vegan diets, despite the carnist responses of veterinarians and some non-government organizations concerned with the welfare of companion dogs. There are well made arguments suggesting that just as we humans might chose a vegetarian or vegan diet for ethical, environmental or health reasons, the same reasons might apply to a meat-free diet for animal companions (Knight, 2015; Peden, 1999; www.vegepets.com). Yet such arguments, and in some cases, experimentation with vegetarian/vegan dog diets, failed to convince the vegetarians and the one vegan in my study that avoiding animal products resulted in an optimal diet for the dog in their care. Whilst the implications of a culture of carnism are clear here ("dogs are essentially wolves"), I also think that the quasi human quality of dogs themselves has a part to play. The normality of pet dogs eating meat means that not feeding animal companions a meat diet raises anxieties about health for those that might otherwise be drawn towards a veg(etari)an diet for their companion animal. The vegetarian interviewees were amongst those who expressed the highest levels of ambivalence about the keeping of dogs as pets and were concerned about their lack of freedom. Whilst contradictory therefore, in this context it is nevertheless understandable that such dog 'owners' would make reference to the preferences or choice that their dogs might make in favor of meat.

In the many posts to be found in response to articles on the web, there are some strongly held views. There are those who express great clarity. This includes the 'vegan view' that: keeping dogs as pets is unnatural, commercial pet food is unnatural, ergo there is no natural diet for a pet dog and dogs can happily be vegan. For others, dogs are carnivores and must be fed as natural a diet as possible, thus raw food, including large quantities of flesh, is optimal. Many are convinced by 'science' and 'expert knowledge' that suggests established 'pet foods' are both the safest and most reliable option, and certainly, the cultural tropes of pet-keeping reinforce this strongly. In the field, and the parks and kitchens where dogs and their people muddle along, there is often ambiguity and a lack of clarity. As such, even those who may contest carnism in their own diet and many aspects of their daily lives, just "don't know about a vegetarian dog". In what Donna Haraway (2008) calls the "sticky knots" of our interactions and co-dependencies with non-human animals, ambiguity is inevitable. Whilst I sympathized more with some of my interviewees' attempts at ethical relating than others, most were very much concerned with trying to live well, or at least to live better, with some other animals. And that rather muddy world of ethical relating is something in which I chose to take heart.

References

Brown, Wendy Y., Vanselow, Barbara, A., Redman, Andrew, J. and Pluske, John R. (2009) 'An experimental meat-free diet maintained haematological characteristics in sprint-racing sled dogs', *British Journal of Nutrition*, 102(9): 1318–1323.

Cudworth, Erika. "'Most Farmers Prefer Blondes'—Dynamics of Anthroparchy in Animals' Becoming Meat." *The Journal for Critical Animal Studies*, 6.1 (2008): 32–45.

Cudworth, Erika. "'The Recipe for Love'? Continuities and Changes in the Sexual Politics of Meat." *The Journal for Critical Animal Studies*, 8.4 (2010): 78–99.

Cudworth, Erika. *Social Lives with Other Animals: Tales of Sex, Death and Love.* Basingstoke: Palgrave, 2011.

Franklin, Adrian. *Animals and Modern Cultures: A Sociology of Human-Animal Relations in Modernity.* London: Sage, 1999.

Fudge, Erica. *Pets*, Stocksfield: Acumen, 2008.

Hall, Tom, Lashua Brett, and Coffey, Amanda. "Stories as Sorties." *Qualitative Research*, 3 (2006): 2–4.

Haraway, Donna. *When Species Meet.* Minneapolis: University of Minnesota Press, 2008.

Iams. "These teeth were made for meat." Accessed July 10, 2014. http://www.youtube.com/watch?v=rT3Ub2Ht_Zo (advertisement uploaded by Iams 24 August, 2012).

Irvine, Leslie. *If You Tame Me: Understanding Our Connections with Animals.* Philadelphia: Temple University Press, 2004.

Jones, Phil, Bunce, Griff, Evans, James, Gibbs, Hannah and Ricketts-Hein, Jane. "Exploring Place and Space with Walking Interviews." *Journal of Research Practice*, 4.2 (2008): 150–161.

Joy, Melanie. *Why We Love Dogs, Eat Pigs and Wear Cows: An Introduction to Carnism.* San Francisco: Conari Press, 2010.

Knight, Andrew. (2005) 'In Defense of Vegetarian Cat Food', *Journal of the American Veterinary Association*, 226(4): 512–513.

Knight, Andrew. (2008) 'Fishy Business?' *Lifescape* (May): 74–76.

Knight, Andrew. (2015) 'Can Cats and Dogs go Vegetarian?', Accessed October 26, 2015. http://www.andrewknight.info/presentations/vegetarian_pets.html

Lau, Edie. "Pet Owners Receive $12.4 Million in Melamine Case." 2011. Accessed June 30, 2014. http://news.vin.com/VINNews.aspx?articleId=20025.

Lavin, Chad. *Eating Anxiety: The Perils of Food Politics.* Minneapolis: University of Minnesota Press, 2013.

Nestle, Marion. *Pet Food Politics: The Chihuahua in the Coal Mine.* London: University of California Press, 2008.

New Zealand Companion Animal Council. "Companion Animals in New Zealand." Accessed March 19, 2015. http://nzcac.org.nz/images/publications/nzcac-canz2011.pdf

Nibert, David. *Animal Rights, Human Rights: Entanglements of Oppression and Liberation.* Lanham: Rowman and Littlefield, 2002.

Peden, James. (1999) *Vegetarian Cats and Dogs*, 3rd edition. Troy, MT.: Harbingers of a New Age.

Pet Food Manufacturers Association. "Pet Population 2014." Accessed July 8, 2014. http://www.pfma.org.uk/pet-population-2014/.

Pet Health Council. "People and Pets." 2008. Accessed April 30, 2010. http://www.pethealthcouncil.co.uk.

Proctor and Gamble. "New Campaign Inspired by Real Stories from Pet Owners." 2012. Accessed July 8, 2014. http://news.pg.com/press-release/pg-corporate-announcements/iams-launches-new-campaign-inspired-real-stories-showcase-e#sthash.FH2P62go.dpuf.

Serpell, James and Paul, Elizabeth. "Pets and the Development of Positive Attitudes Towards Animals." In *Animals and Society: Changing Perspectives*, edited by Aubrey Manning and James Serpell. London: Routledge, 1994.

Torres, Bob. *Making a Killing: The Political Economy of Animal Rights*. Oakland, Ca. AK Press, 2007.

Tuan, Yi-Fu. *Dominance and Affection: The Making of Pets*. New Haven CT.: Yale University Press, 1984.

Vegepets. 'About us' Accessed October 26, 2015. http://www.vegepets.info/index.html.

Wakefield, L. A., Shofer, F. S. and Michel, K. E. (2006) 'Evaluation of Cats fed Vegetarian Diets and Attitudes of their Caregivers', *The Journal of the American Veterinary Medical Association*, 229(1): 70–73.

Williams, Erin and DeMello, Margo. *Why Animals Matter: The Case for Animal Protection*. Amhurst: Prometheus, 2007.

CHAPTER 13

Negotiating Social Relationships in the Transition to Vegan Eating Practices

Richard Twine

With agriculture and especially the commodification of animals contributing significantly to anthropogenic climate change[1]—alongside other key sectors notably transport and energy generation—calls to reduce meat and dairy consumption are now culturally amplified and increasingly mainstream. However these come up against cultural and political obstacles. Firstly politicians or large supermarkets are likely to be timid or captured when it comes to constructing such policy for dietary change and secondly calls for reduced meat consumption are usually made in ignorance of the sort of social complexity inherent to food transitions. This, I argue, gives a role to social science and entails the following research question: how might we *sociologically* begin to approach the challenge of sustainable dietary change? In turn this opens up various potential conceptual approaches and substantive areas of which I focus upon a practice theory approach applied to understanding vegan eating practice.

When compared to omnivorous and vegetarian diets vegan eating practices constitute lower carbon impacts (Berners-Lee et al. 2012; Hoolohan et al. 2013; Scarborough et al. 2014) and are beginning to gain a cultural presence especially in the urban centers of richer nations. As a potential sustainable transition vegan eating represents a practice that on the one hand may be available to significant populations yet is also hindered by the dominance of 'meat culture', what we might call omnivorously normative practices.

Although we could optimistically expect those becoming vegan to be lauded for their compassion and concern for both humans and other animals, transgressing a norm that may itself be constitutive of a particularly dominant narrative of the 'human' entails in most cases, as we shall see, quite a different

---

1 There is a disappointing lack of consensus and peer reviewed work to ascertain the precise contribution of global animal production to overall greenhouse gas emissions. In recent years the United Nations have produced figures of 18% and 14.5% and WorldWatch has claimed 51%. I personally regard none of these figures as wholly reliable which is why I write 'significantly'.

social response. This chapter reports on qualitative semi-structured interviews with 40 vegans carried out between June and December 2013 in three UK cities—Manchester, Glasgow and Lancaster. Vegans constitute an under researched group in UK social science and have not yet figured in research specifically around low carbon sustainability transitions. Although there have been exceptions to a general lack of social science research on veganism (e.g. Jabs et al. 2000; McDonald 2000; Cherry 2006; Sneijder and te Molder 2009; Cole and Morgan 2011; Mendes 2013), it is arguably surprising that there has not been more given the traditional sociological interest in examining minority practice and the transgression of hegemonic social norms.

This study is premised on the assertion that empirical research with vegans is vital to improving our understanding of vegan transition. Listening to vegans talk about their narratives of transition better locates the practice within the ethical, political and relational complexity of everyday life providing lived knowledge that might not emerge from abstract philosophical discussion. This research is intended to contribute to the nascent scholarly consideration of the everyday lived realities of vegan practice (e.g. Potts and Parry 2010). In this chapter I pay special attention to the recursivity between vegan eating practice and social relationships. For example, how do friends and family respond when one first becomes vegan? What types of social negotiation take place both by vegans and those in their immediate social life? I am also interested in how these specific modes of negotiation might *themselves* be active in the normalization of the practice of veganism.

This chapter is broadly divided in two. In the first section I outline the utility of a practice theory approach to the reproduction of (vegan) eating practice and in the second section I use this in order to understand data from my study. Applications of practice theory approaches to sustainable food transitions are rare with the conceptual framework used more typically so far to think about mobility or energy-related practices. I adopt practice theory in order to think sociologically rather than individualistically about vegan eating practice. I am not particularly interested in the view that something like 'attitudes' or 'values' pre-exist or 'cause' practices but instead that particular meanings are recursively bound up in the doing, in the performance of particular practices.

Practice Theory and Eating

Treating practices rather than individuals or institutions as the primary unit of enquiry, considering how they consolidate and change is the focus of practice theory. Here I follow the conceptualization by Shove et al. (2012) of

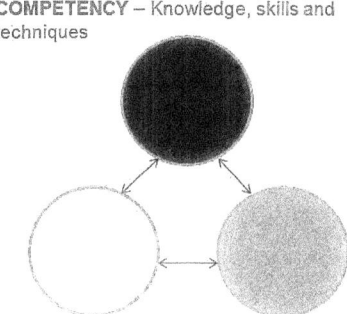

MATERIALS – Things, technologies, infrastructures

MEANINGS – Symbolic meanings, ideas, norms, values, ethics, aspirations

FIGURE 13.1   *Elements of a Practice (Adapted from SHOVE, E et al. (2012) The Dynamics of Social Practice: Everyday Life and How it Changes).*

practices comprised of three elements—*competency, materials* and *meaning*. Competency refers to skills and know-how, materiality to the body and broad array of objects, technologies and infrastructure that comprise a practice, and meanings refer to ideas, aspirations, norms and symbolic meanings (Ibid, 14).

This is a simplification of previous definitions (e.g. see Reckwitz 2002, 249) and perpetuates the practice theory ontological shift that elements are "qualities of a practice in which the single individual participates, not qualities of the individual" (Reckwitz 2002, 250). In this posthumanist inflection people are reframed as 'carriers' of practices (Ibid.). Shove et al. argue that "practices emerge, persist, shift and disappear when *connections* between elements of these three types are made, sustained or broken" (2012, 14–5 original emphasis). This outline of the elements of practice is important because it provides initially a way of potentially thinking about how new practices surface and grow, and how old practices either remain in place or might come to gradually erode. Moreover the dynamism of these connections between elements of a practice is importantly played out through our social relationships and networks that bring us into proximity with new meanings, materials and competences. Thus introducing a focus here on the relationship context of vegan transition is, I contend, especially important in thinking about the capacity of the practice to break through the dominance of meat culture.

All food related practices have historical trajectories and many become socially recognizable and normative. For example in British culture the 'Sunday roast' constitutes one traditional meat eating practice that involves interdependent competences (e.g. preparation, carving), materialities (power, an oven, a tray, a dead bird, a supermarket) and shared meanings (common aesthetic, moral and cultural understandings) that have coalesced to form what practice theorists refer to as the *practice-as-entity* (e.g. Shove et al. 2012, 7) which for many become bound up in family life. The actual doing of the practice in everyday life is referred to as the *practice-as-performance*.

Practice theory wants to acknowledge the corporeality of the social (Reckwitz 2002, 251), it is through such embodied *performance* that a practice in multiple sites is socially reproduced and the connections between the three elements are held in place. Practices then constitute the mundane habits of everyday life and may disappear from both immediate reflection and individual control. This is very much the case with the habitual consumption of animal products that most people do not actively 'choose' but are part of the taken-for-granted norms of culture. Yet it is here also in the performed relations between elements that such links and thus practices may erode, become detached and fade from social life. The case of the Sunday roast for example may in fact be one such practice that is losing practitioners and is no longer as central to weekly routines. Similarly we may contend that the conflations of meat with protein, and of a meal as necessarily including meat, are both meanings increasingly subject to erosion within 'meat cultures'.

As the unit of inquiry practices are intended to cut across traditional sociological demarcations of scale such as micro and macro. Thus Shove et al. (2012) orient themselves to understandings of the extent to which given practices are embedded by arguing that connections form *between* practices themselves, forming what they term 'bundles'. A relevant example here is found in the bundled overlap of meanings in the performance of hegemonic masculinity and the eating of meat perhaps spatially further elaborated during leisure times such as sporting events or barbecues. When such bundles become part of routine social infrastructure we begin to encounter practice 'complexes', more sophisticated, complex, embedded and interdependent practice formations (Shove et al. 2012, 87). Though pre-dating practice theory we might consider the concept of the animal-industrial complex (Noske 1989, Twine 2012) as one such deeply embedded interwoven assemblage of practices that now evokes contemporary concerns over public health, animal ethics and the environment.[2]

Practice theory is a call to understand the dynamism of elements in a practice. Practices are constantly changing although new element configurations can create quite abrupt change. This attention to temporal flow necessarily

---

2 A future aim is to re-write my paper on the animal-industrial complex (2012) using a practice theory approach.

foregrounds the importance of historical analysis for thinking through the biography of elements, practices, bundles or complexes. A more thorough practice approach to veganism than can be presented here requires detailed histories of both veganism and those of its competing practices such as the emergence and growth of meat and dairy production and consumption. Shifting meanings, new materials and new forms of competency change practices and render some redundant, or to use Shove's term, fossilized. Such fossilized elements can become reincorporated into new practices such as a roasting tray being used simply to roast vegetables. Fossils, like sub-cultural or niche elements or practices, are potential resources for thinking about processes of adoption and transition. In addition to changing elements, practices can change when others that they may be bundled together with change first. This introduces different sorts of possible interventions (see Spurling et al. 2013) from those traditionally associated with individualistic approaches to 'behavior change'.

Furthermore, although the primary conceptual focus on practices may suggest a posthumanist decentring of social actors whereby the 'individual' is seen more as a mobile intersection of practices and relationships, our 'doings', our performances, and so our agency working with that of other materialities, the networks we form, remain important for understanding practice transitions. Thus the way in which populations of carriers might change can be important for the normalization or the erosion of a practice (Watson 2012). For example if practitioners become better connected and socially organized they can aid and further embed the practice. We see this with vegan activism toward early forms of normalization of the practice. If the lifecourse or 'career' (Shove et al. 2012) of practices play out via links between elements then traditional sociological foci upon social relations and questions of power (see Sayer 2013, Hargreaves 2011) remain vital to understanding the making and breaking of such links. This is no more obvious than in the continued material and symbolic power of the animal-industrial complex.

In approaching food sociologically and through a practice theory lens we are faced with various conceptual issues. The simple everyday practice of eating intersects with bundles of other practices, notably shopping, transport, storage and cooking practices. Food is part of the materiality of the social itself featuring clearly in the everyday performance of social relationships. Food is also part of doing leisure, doing work, or doing family (Morgan 2011), or performing class, gender or ethnicity. Particular sites such as the workplace or family household are important for exploring food practices. For example the family home has long been seen by social scientists as an important site for social reproduction but translated into the terms of practice theory we can begin to specify how elements of a practice may circulate. If food practices are part of doing family more specifically they can also be seen as part of the performances of childhood, parenthood and grandparenthood and they figure strongly in these negotiated everyday relationships. Sites such as the home must be seen then as significant spaces for the reproduction of meat culture and potentially for the obstruction of vegetarian or vegan practice (see Adams 2001, 133). Family relations are important for the circulation of dispositions, the reproduction of practice and its elements. Parents normalize the regular food *materialities* that a child first encounters and directly or not teach a child various food *competences* such as nutritional knowledge and cooking skills by displaying their own eating performances (Marshall et al. 2007, 168). Food *meanings* are also transmitted via eating performance and familial relations play a role in accomplishing consent for meat culture by negotiating, for example, the contradiction between loving some animals and killing and eating others. Such meanings may be most conflicted in differences between children and grandparents reflecting generationally embedded moral assumptions about certain foods or eating events. Relations are inevitably about affiliation and power and so food practices are also demonstrative of care and affection (Punch et al. 2010, 227) and figure in conflict and control between parents and children (O'Connell and Brannen 2013).

Other relations are generally less immediate and visible; notably the wide array of *production* practices, modes of distribution shaping and shaped by sets of international standards, trade relationships, governance and the political economy of food. Food practices when examined in this way shape the temporalities and spatialities of everyday life. In the area of energy consumption some practice theorists approach energy not as something used for its own sake but as part of accomplishing social practices at home, at work and in moving around. Food arguably is not *quite* so amenable to this framing. Whilst food practices certainly accomplish and lend meaning to a wide array of other practices they also have a sensual aesthetic of their own. This is reflected in the flavor and presentation research of food manufacturers such that the materiality of foods also then has agency in the recruitment of practitioners, constituting part of the social, importantly defined in more than human terms. This broadening out of the social to include a consideration of materialities as an important element of practice reflects developments in social theory acknowledging the way that "social relations are 'congealed' in the hardware of daily life" (Shove et al. 2012, 10).

Doing Vegan Research

In ongoing work I develop a fuller conceptualization of a practice theory approach to vegan transition that involves a closer examination of the specific competences, materials and meanings that comprise vegan eating practices drawing upon the same data used in this study. That work also draws more explicitly upon the insights of practice theory for thinking how practices might change with the aim that the approach can add much to theorizing the scaling up of vegan practice and the erosion of meat culture.

The everyday performance of vegan eating practice via close connections between elements has already coalesced into a do-able practice for millions of practitioners across Europe, Australasia and North America.[3] There are clearly important questions here that I only touch upon in this chapter around dispositions and practice (Southerton 2013, 16) such as the gendered, classed and racialized dimensions of vegan practice, but are nevertheless reflected in the narrow demographic of my sample. For those vegans in my study the main threat to, or difficulty within, practice performances was found not so much in the access to practice elements but in the social, relational responses to veganism encountered. Thus in effect my participants represented those practitioners who were, at least at the time of interview, successfully negotiating an at times difficult social environment. Successful socially competent vegan practitioners represent a valuable resource for thinking about forms of negotiation in the everyday performance of doing vegan eating.

As well as living in a 'meat culture' we also live in a male dominated 'car culture' and both practices of eating meat and driving share some meanings with the hegemonic performance of masculinity. However food practices are also uniquely relational in comparison to other sustainability transitions and meat/dairy consumption is specifically potent within the very meanings of 'performing human' vis-à-vis other practices. If one decides to give up driving and instead adopt cycling and public transport practices it would be difficult to anticipate too much in the way of a negative social response. As we shall see exiting from meat culture, from the practices of animal consumption—in this case food—evoke a markedly different and perhaps unique response in many cases. It seems to really 'touch a nerve'.

---

3   There are no known reliable figures for numbers of vegans in these geographical areas. An estimate for the UK would be in the low hundreds of thousands. There are of course gray areas in the definition.

Participants in this study were made aware in advance of my own veganism which was intended as a positive disclosure that established rapport. This inevitably introduced elements of auto-ethnography into the study as I reflected back to my own personal narrative those of the participants. It is likely the case that in order to come to fruition research on vegans and veganism relies upon the *interestedness* of vegan researchers in a manner similar to engaged feminist research. Epistemologically all research is 'interested' and so the approach favored here in approaching veganism is to 'frontstage' the reasons for doing so.

As a vegan sociologist located within the field of Critical Animal Studies (CAS) and climate change research this study is informed by the evidence base cited at the outset that vegan eating practice can constitute the lowest carbon footprint. Furthermore this research is embedded in attempts to contest, provoke and dismantle the animal-industrial complex based as it is within systems of capital accumulation at the expense of sentient animal (including human) forms of life. Meat culture is thus based upon an anti-scientific and dated disavowal of animal subjectivities. The scale and intensification of animal exploitation has increased at the very same time that scientific knowledge of the sentience of other animals has emerged. This is a moral contradiction that asks pivotal questions of what it is we understand ourselves to be when we describe ourselves as 'human', but more specifically questioning of the way in which the exploitation of animals became so successfully bound up in twentieth century projects of capitalist growth. Vegan practice exposes the irrationality of killing animals for food in most locations[4] and asks questions of a delirious and escalating trajectory of exploitation that Derrida described as "no longer knowing where it is going" (2008, 25).

Even if we were to somehow subtract this engaged and pro-vegan CAS framing there would remain a self-interested rationale for the normative omnivorous majority to also support research into veganism. As calls for a more protectionist and palatable discourse of 'meat reduction' coalesce in the context of climate change mitigation, vegan practice and competency becomes important generally for those times and days when the omnivore has decided not to eat meat. Thus shifting the centrality of the norm of animal consumption is important even in the more conservative discursive practice of meat and dairy *reduction*.

---

4 For a recent essay on veganism and universalism (see Twine 2014).

Method

Between June and December 2013 forty vegans in three UK cities—Manchester (14), Glasgow (14) and Lancaster (12) were interviewed. After University research ethics approval was obtained participants were recruited initially through an advert in the magazine of the Vegan Society, through local vegan organizations and word of mouth. Once a certain momentum was reached the sample was simple to obtain via the snowball technique. The interviews were semi-structured and open but were guided by three main question areas. First participants were asked to narrate their own story of transition to veganism. Secondly participants were asked about their everyday doing of veganism which was approached by asking questions about the materials, competences and meanings of the practice; as well as participants' involvement in forms of vegan social organization. Finally participants were asked a set of questions about transition and relationships. It is this latter area that I focus upon in this chapter. Interviews lasted between 40 and 75 minutes and took place either at the participant's home, my home, in my office or in a vegan friendly cafe.

Participants were aged between 18 and 72 years, with the average age 36.8 years. Twenty-nine participants (73%) were female. Thirty-nine participants (97%) were self-defined white British, one was self-defined mixed race British. Thirty-one participants (77%) either had a first degree or were studying for one. Although this study makes no pretence toward a representative sample and is instead focused upon what can be learnt from rich in-depth accounts the sample did reflect common assumptions about the broader vegan community as disproportionately female, educated and white. Clearly if only a narrow demographic is disposed toward and captured by the practice then that may be instructive for examining the meanings of veganism and those of animal consuming practices. Such analyses can potentially inform strategies of vegan mainstreaming.

The areas of interview data I shall focus upon here were in response to questions asked around social relationships and vegan transition. These included asking participants 'Can you remember what sort of reactions you experienced from those close to you?', 'Did any friends or family disapprove of you becoming vegan?', 'Does your veganism become a problem when you eat with your family?' and 'Has your veganism been a problem with partners/close relationships?'. In some cases answers to these questions arose earlier in the interview when participants were narrating their transition stories. In presenting results from this study here I focus in only upon two related themes that emerged from the data. Firstly, initial reactions from friends and family and, secondly, modes of negotiation between vegans and omnivores.

Becoming Vegan in a Normatively Omnivorous Meat Culture

A majority of participants in this study reported negative reactions from friends and family in their decision to become vegan. Thirty-three of forty participants (82.5%) reported at least some examples of negative reaction. This tended to dissipate over time, a point which I shall return to. Some had a support base of vegan friends and a smaller number came from families with some history of vegan or vegetarian practice. A small minority went vegan with friends which resulted in a more supported transition. However in the majority negative response we can see a lack of support for the positive meanings of veganism and a heightened need for social and emotional competency for those becoming vegan. This negative response materialized in various ways which I now review. The following represent articulations from participants of responses from friends and family to vegan transition. Only a couple of participants had transitioned directly from omnivore to vegan so some answers also include discussion about the first transition to vegetarianism. All names are pseudonyms.

> ... people just don't get it
> —JOHN

• • •

> I think with family there's more confusion, it had taken them a long time to understand what vegetarianism was and my sister isn't happy because the only thing that she would ever cook is cheese and onion pie and now she's like, what can I cook you?
> —LAURA

• • •

> ... over the years I've had a lot of mickey taking
> —REBECCA

• • •

> ... there were a couple of occasions early on where I didn't get invited to family events because they said, well we only will have meat, so you can't come
> —FIONA

• • •

You know what is interesting; I didn't actually get an honest response from my best friend. She just kind of went, oh right ok. And I could tell she wasn't happy and she said, oh well you be careful what you're eating sort of thing and then it was a bit of a closed book and she didn't really want to talk about it and she didn't want to know why and I found out that she'd gone straight to my wife with all of her concerns and all of her questions and all of her, oh my God, what the heck is Natalie doing and all of this business
—NATALIE

• • •

I remember my gran saying, cos obviously I went vegetarian, and I wasn't seeing her very often because I was away at university and she said, oh so you'll only eat egg sandwiches, and now you won't even eat the egg sandwich, 'what am I going to put in your sandwich?'
—KATE

• • •

Family's not very supportive of it because none of them are even vegetarians so I don't think they really understand. My mum just thinks I'm too extreme
—BRIDGET

• • •

My mother was horrified, my nana was horrified and a couple of my friends were horrified and my mum just refused to cook for me. She's probably cooked for me about four times in the last five years
—JANET

• • •

My big sister thinks the whole thing is ridiculous
—SUZANNE

• • •

A couple of my friends sneered at me
—SOPHIE

• • •

My mum would cook, but because she's vegetarian and my dad eats meat so she told me straight out that she wasn't prepared to let me be vegan while I was in the house cos it would mean cooking three separate meals and she didn't want to lose that family time of having a sit down meal together
—ANNIE

• • •

Friends, well a couple were just like why, just why? Some I think, if ever I talked about it or whatever I think they would feel threatened, as though, oh I don't know how to explain it. Like it was an attack on them if I was to, like a personal attack if I was to talk about it
—CARRIE

• • •

My mum said 'don't forget we do need to eat meat, to get vitamins'
—DON

• • •

When I went veggie my parents went completely up the wall about it. Went I went vegan three years ago, my girlfriend at the time she was completely anti it. That played a role in the relationship not lasting
—PATRICK

• • •

Family totally took the mickey and they were actually quite nasty at times. And they still say stuff like "just have a steak, just have a steak". They are not very sensitive at all. For a while they thought it was some kind of hipster thing that I was doing, some sort of trend. And they thought 'oh, you'll be over it in a week'.
—BOB

• • •

I don't recall getting massively positive responses. Certainly my husband at the time was just like 'Oh no, no, no, no, no, no' (laughs), to the point where he was just like 'please just be vegetarian'. Whereas if I'd said I was going to be vegetarian, he would have been like 'no, no, no, no, no'. My mum was quite concerned when I eventually told her
—JOANNA

• • •

My granny even after 5 years thinks I'm insane, because she was brought up on this traditional diet. It's not really within her sphere of thought why would you want to live without eating meat
—TANYA

• • •

My dad told me I was being ridiculous
—DIANE

• • •

My dad said 'you're spoiling yourself, why are you denying yourself all these things?'
—MARY

• • •

My mum was probably the person who had the biggest over reaction ever. She was really concerned about my health. She was really worried that I was going to waste away
—NICOLA

• • •

My best mate told me that he'd never cook for me again, that I was being really ridiculous, over the top, couldn't believe that I was doing it. I'd never be invited to dinner parties again. And just went mental about it. He just thought it was like massively extreme
—EMILY

• • •

> Not disapproving, but not understanding. Their (friends) understanding of it is, we're meant to eat meat
> —ANTHONY

• • •

> Most people were quite negative, constantly asking why and that was just annoying after so many times... Mostly school friends were shocked, they used to shove ham sandwiches in my face which was just odd
> —ALEXANDRA

⋮

These extracts represent much of the negative social response that the majority of participants experienced when becoming vegan. It is potentially significant that so many of the participants did encounter this kind of response. If this is typical for people becoming vegan or trying to become vegan the interdependency between the practice and relationships could be an especially important site for examining the tension between meat culture and a counter hegemonic practice like veganism.

Framed through a practice theory lens that sees strong connections between competences, materials and meanings as necessary for the reproduction of practice we can note in this sample a distinct lack of know-how of vegan practice (nutritional knowledge, knowledge of, and how to cook, vegan meals) amongst the friends and family of those who themselves were trying to become competent in the everyday doing of veganism. A lack of competency in those close to vegans in turn impacts upon the availability of the right materialities (vegan foodstuffs and knowing where to buy them) especially where responsibilities in the home for shopping and cooking are shared with omnivores. Furthermore responses from significant others that devalue vegan practice introduce a sense of failed reciprocity around the positive meanings of veganism that have been important in drawing people to the practice. Above we note people as 'not getting it' and 'confusion' and I would suggest this can refer to all three elements in the practice but certainly to a lack of empathy over the affective lure and meaningfulness of performing veganism. It is not difficult to see how this social environment could mitigate against the uptake of vegan practitioners. In effect this introduces the need for a form of additional competency that involves skills of emotional and social negotiation that would be less necessary in a more supporting environment.

Further responses above can be organized thematically around the omnivore's defensiveness and a construction of the vegan as awkward. Vegans introduce a sense of embodied questioning, a discomfort to the habitual normativity of meat culture. In parallel to this there is a response that could be read as arising out of defensiveness over the implicit questioning of animal consuming practices. In a sense the violence that meat culture more or less successfully sequesters to the spatial and class margins of society is brought back into uncomfortable proximity by the presence of the vegan. Meat/dairy practitioners then work to defend their practice through uses of humor, 'mickey taking', and in labeling vegan practice negatively as 'extreme' as in some of the extracts above. It is also arguably possible to place within this theme of defensiveness particular discursive strategies of naturalization that seek to consolidate the normativity of meat culture. These are sociologically interesting moments because the taken for granted practice of meat/dairy consumption is forced to bring what are often its unreflected meanings into play. We can see this above in Don's mother's view that 'we do need to eat meat, to get vitamins' and in Anthony's friends view that 'we're meant to eat meat'. These are moments in which animal consuming practices rehearse and reproduce their meanings in competitive tension with vegan practice. Such statements evoke an essentialism of the human and place meat culture as part of an imagined fixed natural order rather than a political, economic and social norm. Less blatant but related to these forms are expressed concerns for health received from friends and relatives. This is not to deny genuine relations of care and the importance of nutritional competency for vegans and also for the wider population. Naturalization strategies also seem able to ignore the evidence of healthy vegan performances which undermine their own credibility.[5]

The vegan as awkward is a further theme that emerges from the responses above. This also relates to the potential lack of vegan competency amongst omnivores in that either it is not known what to cook for vegans as in Laura's extract, or cooking an extra dish might be seen as impractical and time consuming as with Annie who also introduces the point of a threat to commensality and being controlled by her parent. Although the awkwardness of veganism is a product of the normativity of meat culture it would make little sense to dismiss it for that reason. In social situations where vegans live with omnivores

---

5 Vegans of course sometimes use their own strategic essentialism in order to argue that humans are 'naturally' vegan.

there is a social disruption such as a threat to eating together not just due to practical reasons but because of the way in which vegans begin to see animal consumption. For many vegans, including many in my sample, eating alongside omnivores becomes affectively difficult. Becoming vegan can create social distance which can change the quality of relationships and in some circumstances include the breaking of some relationships and the seeking out and forming of new relationships with other vegans. It can shape social exclusion as well as self-imposed exclusion. As well as a good example of the recursivity of practices and relationships these relational dimensions deserve further research since they can be crucial to the everyday performance of veganism. In contrast to the extracts above there were examples in my sample where the omnivore's response was notably more open and positive and most of those negative responses were seen to adapt, change and accommodate over time.

Non-Practicing Practitioners

Whilst there is a lot more to be teased out from the data, due to space constraints I will finish with two examples of social negotiation that emerged from my sample. These act as forms of bridge-building and serve to lessen social distance between vegans and others and, importantly, I argue, further embed and normalize vegan practice. In spite of the negative initial reaction most participants reported that the situation improved over time. A noticeable theme is the emergence of what I term 'non-practicing practitioners'. Although this sounds oxymoronic I suggest that friends and relations of vegans are more likely to become 'non-vegan vegan advocates' who inadvertently promote the practice. As a demonstration of care, friends and family members will often start to demonstrate certain aspects of vegan practice. Several examples of this emerged from participants.

> It's great now because my dad cooks a lot of vegan food anyway
> —FIONA

• • •

> …they appreciate it a lot more now and now my mum will go out of her way, I think she kind of enjoys it in the shop, she'll come back and she'll have bought me a vegan yoghurt or something or different kinds of chocolate milk and stuff and I think she kind of enjoys that now
> —ANNIE

• • •

…she (mother) tries to experiment with vegan baking now, she tries to adapt recipes so they'll be vegan or try out vegan recipes. Yeh she's embraced it a lot more, in fact I think her eating habits have changed as a result
—CARRIE

• • •

After I'd told him (partner) of my decision whilst on holiday, he made me the most fantastic dessert on my return, this chocolate rice pudding dessert, totally vegan
—LESLIE

• • •

…next week I'm going down to meet them (parents), and I'm meeting them a bit away from where we'd usually meet. And my dad's been onto the internet checked out the restaurants, and phoned up one of them
—LESLIE

• • •

My mother is a very good baker, and she always makes something for me when I visit. She no doubt has to adjust her style for me, so I probably set an example in a way. She has expanded her repertoire as it were. She exactly now thinks that the vegan margarine alternatives are as good if not better
—PAUL

• • •

…she (mother) was quite interested in how easy it was and she's been really really supportive and I'm going to stay with her next week and she's researched all these vegan things and got me in all this stuff to eat which is fantastic and lovely
—NATALIE

• •
•

These extracts represent examples of non-practicing practitioners who support the veganism of those close to them. Although not (yet) vegan themselves they are nevertheless reproducing elements of the practice. Specifically they have developed forms of *competency* in vegan cooking, the sourcing of vegan food, the definition of vegan food, engaging with non-vegan restaurants and in eating vegan food more of the time. They are clearly also then engaging with parts of the *materiality* of vegan practice. In practice theory terms they are not making strong links between these elements and the *meanings* of veganism which is one explanation of why they have not become vegan themselves. However relations of care and affinity draw them partly into the practice and they provide a peripheral, ambassadorial support for the normalization of the practice. This could be important for practice theory to note that the boundary of those within a community of practice is not wholly clear and peripheral non-practicing practitioners may also be important to the trajectory of a practice.

Demonstrative Vegan Practice

A second form of social negotiation which also builds bridges and this time comes from vegans themselves pertains to the mode of performing veganism in a demonstrative manner that draws omnivores or vegetarians into the sensual experience of vegan food. The example of vegans cooking for non-vegans arose several times in the sample and had the effect of bringing vegan eating into a familiar space and as a known, appreciated aesthetic experience again expressed via relations of care. Here are just two examples:

> They still ask (why I'm vegan) but because they saw me cooking that really helped. Even after a year I'm still not used to cooking for one person, so if I make a meal it would usually be enough to feed 3–4, so people would always eat my food. And then they'd be like 'Ah Alexandra, you make such amazing food'. That's really good because they know vegan eating isn't something strange or horrible
> —ALEXANDRA

• • •

> I think what helped enormously was me cooking for her (mother), I think that helped massively and I just had to bite my tongue and invite her round for dinner and make like a big three course meal. She was very

surprised that it actually tasted nice, very surprised and especially when I made cakes and things and eventually she became quite proud of me doing that

—JANET

∴

This mode of performing veganism with non-vegans which is sometimes scaled up in the example of the vegan potluck can serve to restore a sense of commensality and social connection with food. Here a vegan is able to demonstrate their cooking competency and skill and the material agency of the food is, in a sense, allowed to do the talking. In her book *Living Among Meat Eaters* (2001) Carol J. Adams refers to such demonstrative veganism as akin to the work of a magician in that the vegan cook 'shows but does not tell' (196). To tell would be to open the door to the likely negative meanings an omnivore has of vegan food (materiality). In such re-crafted practice eating assumes familiar form via vegan experimentality, creativity or substitution and the meanings of veganism are allowed to transition from near pathological to a new normal.

Concluding Remarks

These two examples of bridge building and negotiation contrast markedly with the narrated negativity that the vegans in my sample reported during the early days of becoming vegan. They suggest that normalization of the practice can take place within the very strategies of social negotiation adopted by vegan practitioners and those close to them. This ought to be a useful finding for those interested in expanding the scale of vegan practice and for imparting advice to new vegans or the 'vegan curious'. Particular ways of doing veganism may be more successful than others in circulating the practice. The more circulation the less likely that new vegans will be confronted with the sort of negative responses typical of my sample during initial transition.

Further analyses of this data and new studies can undoubtedly add to the sociological analysis of lived everyday vegan practice. This provides an alternative to individualistic, behavioral and some philosophical approaches that tend to view the agent of change as the individual rather than practices. Such approaches still tend to be assumed by most forms of animal advocacy. One limitation of such approaches is to over focus on certain meanings (values, ethics) and to extract these both from the everyday lived performance of (eating) practices and from their interrelationship with forms of competency and materiality. Simply imparting 'better ethics' to people ignores how everyday life is caught up in relational complexity and is reproduced through *routinized* forms of practice. In order to change routines all elements of a practice must be brought under scrutiny.

This chapter turned to practice theory as a promising frame for understanding the sustainable food transition of veganism. It proved useful for analyzing the recursivity between relationships and food practices and offers distinctive strategies for intervention that I turn to in concurrent and future work. Vegan eating practices offer a tantalizing lower carbon alternative that additionally contests the negation of animal life in meat culture. Since human/animal dualism reproduced, for example, in performances of animal consumption is such a potent meaning within hegemonic reiterations of the human, veganism faces unique challenges as a practice in search of practitioners. Such challenges play out in the everyday relational struggles between novel vegan practice and those performing the work of the animal-industrial complex.

References

Adams, Carol. J. *Living Among Meat Eaters: The Vegetarian's Survival Handbook* New York: Continuum, 2001.

Berners-Lee, M. et al. "The Relative Greenhouse Gas Impacts of Realistic Dietary Choices." *Energy Policy* 43 (2012): 184–190.

Cherry, Elizabeth. "Veganism as a Cultural Movement: A Relational Approach." *Social Movement Studies* 5.2 (2006): 155–170.

Cole, Matthew, and Karen Morgan. "Vegaphobia: Derogatory Discourses of Veganism and the Reproduction of Speciesism in UK National Newspapers." *The British Journal of Sociology* 62.1 (2011): 134–153.

Derrida, Jacques. *The Animal that Therefore I Am*. New York: Fordham University Press, 2008.

Hargreaves, T. "Practice-ing Behaviour Change: Applying Social Practice Theory to Pro-environmental Behaviour Change." *Journal of Consumer Culture* 11.1 (2011): 79.

Hoolohan, Claire et al. "Mitigating the Greenhouse Gas Emissions Embodied in Food through Realistic Consumer Choices." *Energy Policy* 63 (2013): 1065–1074.

Jabs, Jennifer et al. "Managing Vegetarianism: Identities, Norms and Interactions." *Ecology of Food and Nutrition* 39.5 (2000): 375–394.

Marshall, David et al. "Families, Food, and Pester Power: Beyond the Blame Game?" *Journal of Consumer Behaviour* 6.4 (2007): 164–181.

McDonald, Barbara. "'Once You Know Something, You Can't Not Know It' An Empirical Look at Becoming Vegan." *Society & Animals* 8.1 (2000).

Mendes, Elisabeth. "An Application of the Transtheoretical Model to Becoming Vegan." *Social Work in Public Health* 28.2 (2013): 142–149.

Morgan, David H. J. *Rethinking Family Practices*. Palgrave Macmillan, 2011.

Noske, Barbara. *Humans and Other Animals: Beyond the Boundaries of Anthropology*. Pluto Press London, 1989.

O'Connell, Rebecca, and Julia Brannen. "Children's Food, Power and Control: Negotiations in Families with Younger Children in England." *Childhood* 21.1 (2014): 87–102.

Potts, Annie, and Jovian Parry. "Vegan Sexuality: Challenging Heteronormative Masculinity through Meat-free Sex." *Feminism & Psychology* 20.1 (2010): 53–72.

Punch, Samantha et al. "Children's Food Practices in Families and Institutions." *Children's Geographies* 8.3 (2010): 227–232.

Reckwitz, Andreas. "Toward a Theory of Social Practices." *European Journal of Social Theory* 5.2 (2002): 243–263.

Sayer, Andrew. "Power, Sustainability and Well Being: An Outsider's View." In *Sustainable Practices—Social Theory and Climate Change*, edited by E. Shove and N. Spurling. London, Routledge, 2013.

Scarborough, Peter et al. "Dietary Greenhouse Gas Emissions of Meat-Eaters, Fish-Eaters, Vegetarians and Vegans in the UK." *Climatic Change*. Online only as of June 2014: 1–14.

Shove, Elizabeth, Pantzar, Mika, and Watson, Matt. *The Dynamics of Social Practice—Everyday Life and How it Changes*. London, Sage, 2012.

Sneijder, Petra, and Hedwig te Molder. "Normalizing Ideological Food Choice and Eating Practices. Identity Work in Online Discussions on Veganism." *Appetite* 52.3 (2009): 621–630.

Southerton, Dale. "Habits, Routines and Temporalities of Consumption: From Individual Behaviours to the Reproduction of Everyday Practices." *Time & Society* 22.3 (2013): 335–355.

Spurling, Nicola et al. Interventions in Practice: Re-Framing Policy Approaches to Consumer Behaviour. 2013.

Twine, Richard. "Revealing the 'Animal-Industrial Complex': A Concept and Method for Critical Animal Studies." *Journal for Critical Animal Studies* 10.1 (2012): 12–39.

Twine, Richard. "Ecofeminism and Veganism—Revisiting the Question of Universalism." In *Ecofeminism: Feminist Intersections with Other Animals and the Earth*, edited by Carol J. Adams and Lori Gruen. London, Bloomsbury, 2014.

Watson, Matt. "How Theories of Practice Can Inform Transition to a Decarbonised Transport System." *Journal of Transport Geography* 24 (2012): 488–496.

CHAPTER 14

Critical Ecofeminism: Interrogating 'Meat,' 'Species,' and 'Plant'

Greta Gaard

> Do we need a 'plant ethics' that responds to vegetal instrumentalization in an allied manner to the ways animal ethics has responded to the animal-industrial complex?
> —JONI ADAMSON AND CATE SANDILANDS, Vegetal Ecocriticism (2013)

∴

The above question might alert animal studies scholars to a pending omnivore backlash against the decade-long success of animal studies, launched by Jacques Derrida's "The Animal That Therefore I Am" (2002) which catapulted vegan perspectives into academic credibility. Like animal studies, plant studies scholarship has been ongoing for some time, but only recently has emerged as a cutting-edge academic field. One could compare Peter Singer's *Animal Liberation* (1975) with Peter Tompkins and Christopher Bird's *The Secret Life of Plants* (1973) as foundational works for two movements that would later become recognized as companion branches of posthumanist thought, though in thankless academic fashion, the founders of each branch have been strongly critiqued, Singer for his human rights-based moral extensionism, and Tompkins and Bird for their 'new age' unscientific speculations. Invoking the field of "critical plant studies" and author of *The Omnivore's Dilemma* (2006), the carnivorous locavore Michael Pollan, Adamson and Sandilands anticipate my comparison in their description of a "Vegetal Ecocriticism" pre-conference seminar for the Association for the Study of Literature and Environment (ASLE) 2013 biannual conference:

> As critical animal studies and animal rights scholars/activists have effectively worried constitutive boundaries between human beings and other animals, plant studies scholars have questioned the similarly political line between plants and animals: plants communicate, move, decide,

transform, and transgress in ways that are sometimes uncomfortably 'like' animals (including humans), and sometimes so completely Other to animality that conventional metaphysical principles are radically denaturalized.

The phrase "similarly political line between plants and animals" might raise eyebrows among animal studies scholars, and understandably so. Vegan and vegetarian feminist ecocriticism has a substantial history, starting with Carol Adams' discussion of Frankenstein's vegetarian monster in *The Sexual Politics of Meat* (1990), and continuing through the work of many vegan and vegetarian feminist ecocritics (i.e. Armbruster 1998; Chang 2009; Donovan 1991, 2009; Gaard 2000, 2001, 2013). But Simon Estok's (2009) essay in the flagship journal of the Association for the Study of Literature and the Environment (ASLE) finally threw down the gauntlet. Defining 'ecophobia' as an "irrational and groundless hatred of the natural world," Estok argued that "ecophobia is rooted in and dependent on anthropocentric arrogance and speciesism"; thus, it is "difficult to take seriously... the ecocritic who theorizes brilliantly on a stomach full of roast beef on rye" (2009, 208, 216–17). Four years after placing carnivorous ecocritics under such scrutiny, the emergence of Randy Laist's *Plants and Literature: Essays in Critical Plant Studies* (2013) confirms that vegan/vegetarian ecocritics will soon need a response to 'vegetal ecocriticism,' just as animal studies scholars will have to consider the findings and claims of plant studies. So, what is 'plant studies,' and do its claims of animal/plant similarity seek to delegitimate the very real suffering of other animal species, and place human food choices on a terrain of moral relativism suitable to a carnist[1] culture?

Central to these questions are the key terms *meat* and *species*.

Plant Studies: A New Field Emerges

Although articles formulating the emergence of plant studies had already begun to appear in journals such as *Quanta* (McGowan 2013), *Mother Earth News* (Angier 2013), *Journal of the Fantastic in the Arts* (Miller 2012), *Societies*

---

[1] Melanie Joy (2010) uses *carnism* to describe cultures that make meat-eating seem "normal, natural, and necessary"—effectively, hegemonic—through a conceptual schema that uses "objectification, deindividualization, and dichotomization" (96, 117). Joy's work popularizes concepts from more sophisticated philosophical arguments such as Val Plumwood's (1993) Master Model and the wealth of groundwork provided by vegan and vegetarian scholar-activists.

(Gagliano 2013; Ryan 2012), and even the *Journal for Critical Animal Studies* (Houle 2011), plant studies emerged into popular culture through the publication of Michael Pollan's (2013) "The Intelligent Plant" in *The New Yorker*. There, Pollan reports the findings of biologists—molecular, cell, plant—confirming capacities that new materialists call *agency* (Coole and Frost 2010) and suggesting paradigm-shifting parallels to animal capacities as well (see Table 1).

TABLE 1   *Behaviors confirming agency in plants and animals**

Species / Behaviors confirming agency	Plants	Animals
Senses: apparatus and ability to sense and optimally respond to environmental variables	15–20 distinct senses: • sense and respond to chemicals in the air or on their bodies • react differently to various wavelengths of light and shadow • a vine or root 'knows' when it encounters a solid object, and vines grow toward supports • plant behavior suggests plants hear the sound of flowing water, and respond to potential threats by generating defense chemicals	5 senses: • Smell • Taste • Sight • Touch • Sound
Communication	Plant 'signaling' occurs through the release of volatile chemicals, or the production of predator-repelling toxins	Vocalizations, body movements, postures, scents
'Intelligence'	Plants know their environment, location, and other plants nearby	Brain, neurons, nervous system, consciousness
	Root tips gather and assess data from their environment and respond in ways that benefit the plant community, kin and beyond	Ability to reason, judge
	Store information biologically, through molecular wrapping around chromosomes (epigenetics)	Memory & learning: laying down new connections in a network of neurons

Species / Behaviors confirming agency	Plants	Animals
Self-identity	'Distributed intelligence' through root networks; know their environment; may use fungal networks to nourish seedlings and even trade nutrients across subspecies	Self-awareness as individuals, family and species members
Arguments for uniqueness (and hence, moral standing, moral consideration, and possibly 'rights')	Can lose up to 90% of their bodies without being killed Plant signaling: molecular 'vocabulary' releases to signal distress, deter or poison enemies, recruit animals to perform services (i.e. pollination)	Ability to feel emotions, i.e., love, anger, loyalty, joy, playfulness, grief, depression, appreciation of beauty, loneliness, compassion, jealousy, regret, sociality

* Data sources from Angier 2013; Bekoff & Goodall 2007; Bekoff & Pearce 2009; Chamovitz 2012; Gagliano 2013; Pollan 2013; Ryan 2012.

Not surprisingly, there are at least two dissenting branches of the field. Anxious to safeguard their work as legitimate science by avoiding anthropocentrism and animism, conservative (read 'hard science') plant scientists interpret their data in very humanist ways that preserve the animal/plant species hierarchy, rejecting the terms 'plant communication' and 'plant neurobiology' for 'plant signaling,' and 'learning' for 'adaptation' (Pollan 2013). More progressive plant studies scholars (read 'humanities'), however, suggest not only that we should "stop anthropomorphizing plants" but actually "try instead to think like them, to phytomorphize ourselves" (McGowan 2013). Challenging evolutionary biology's misuse of the concept, 'survival of the fittest,' Monica Gagliano concludes that even "the very competitive evolutionary process of natural selection involves cooperation," and "cooperation and competition can coexist" because among plants, more cooperative, "collective associations are indeed an ecologically common state of affairs" (2013, 153). "Thinking plant-thought shoves us in a better way than thinking animal-thoughts does," argues Karen Houle in the *Journal of Critical Animal Studies*, since the ecologically "'correct unit' of analysis is not the individual, nor the dyad, but 'the assemblage'" (2011, 111).

Though not explicitly drawing on queer studies, Houle's argument uses a posthumanist methodology that compares favorably with queer methodology (i.e., Browne and Nash 2010). As Cary Wolfe (2009) explains, posthumanism involves both content and method:

> ...one can engage in a humanist or a posthumanist practice of a discipline.... Just because a historian or literary critic devotes attention to the topic or theme of nonhuman animals doesn't mean that a familiar form of humanism isn't being maintained through internal disciplinary practices that rely on a specific schema of the knowing subject and the kind of knowledge he or she can have. So even though your external disciplinarity is posthumanist in taking seriously the existence and ethical stakes of nonhuman beings (in that sense, it questions anthropocentrism) your internal disciplinarity may remain humanist to the core. (572)

Both posthumanist and queer methodologies reject the essentialist, unified Cartesian human for a socially constructed plurality of continually shifting identities and selves;[2] both methodologies can be seen in Houle's approach to plant studies. For example, Houle provocatively rejects the humanist and heteronormative hegemony of mutualism as a framework in plant studies, challenging the conceptual gesture that defines plant behaviors as 'communication' only in mutualist dyads: "if the benefits to the emitter and receiver [of plant signals] are not equal and not mutual, the description of the plant behavior is downgraded from 'communication' to 'eavesdropping'" and the "third party is called a 'cheater'" (109). Instead, Houle suggests framing these communications not as 'illicit' but as "actions of generosity and gift...spontaneous, non-meritocratic...uncontainable excess" (109). Pointing to the "permanent and varied role of organic and inorganic thirds and fourths in every communication mechanism" (110), Houle invokes Deleuze and Guattari's

---

2 As queer studies scholars have argued, queer perspectives may endorse but are more likely to differ from the liberal assimilationist goals of the lesbian/gay/bisexual/transgendered (LGBT) movement for inclusion in heteronormative institutions (i.e., equal marriage, equal inclusion in the military, LGBT human rights legislation, corporate sponsorships for Pride). Present in groups such as the Radical Faeries of the 1960s and queer activist groups of the 1990s, ACT UP and QUEER NATION, some queers have expressed resistance to heteronormative assimilation, choosing instead to celebrate queer culture, eschew essentialist dualisms of gender and sexuality, and affirm queer diversities across race, gender, and class (Jagose 1996; Gleig 2012). These perspectives form the base for queer methodology (Browne and Nash 2010).

concept of 'unholy alliances' to describe plant relations as "a radical collectivity" that transforms sociality and kinship "beyond any simple sense of between" to a broader "among" (111). Houle advocates 'becoming-plant' for the ways it "opens up thinking about relations as transient alliances rather than strategies," and "credits the accomplishment of identity and intimacy as *a radically collective achievement*" (112). These arguments fit well with new materialism's concept of *transcorporeality* (Alaimo 2010) as well as queer theory's fluidity of identity, sexuality, and community.

Leading ecocriticism's vegetal branch of 'critical plant studies,' Catriona Sandilands (2014) conceives of plant studies as emerging from and companioning critical animal studies. Her work on *Queer Ecology* (Mortimer-Sandilands and Erickson 2010) has explored the ubiquitous presence of queer animals that usefully complicates heteronormative assumptions about sexuality, embodiment, and authenticity. Following models of queering animal studies, Sandilands (2014) proposes the concept of 'botanical queers' that illuminates how plant lives offer the potential to complicate heteronormative (and humanist) conceptions of identity, kinship, and time. Though Sandilands' version of critical plant studies seems to propose a sibling relation with animal studies, other approaches may be less benign.

For example, Canadian science writer Elaine Dewar (2013) seems almost gleeful in speculating that the proposal that 'plants think' will be "anathema to vegans," and since chemistry professor Susan Murch has called the volatile chemical signals from wounded plants "screams," Dewar warns, humans should "remember that, the next time you rip a carrot out of the garden." It appears that "a specter is haunting animal studies," writes T. S. Miller, "the specter of cellulose" (2012, 460). Miller rightly criticizes the humanist methodology of animal studies, arguing that "it is zoocentrism and not simply anthropocentrism, the bugbear of animal studies," that defines human identity (463). Miller agrees with Matthew Hall's view that "zoocentrism helps to maintain human notions of superiority over the plant kingdom in order that plants may be dominated. It is a crucial dualising force, responsible for depicting plants as inferior beings and as the natural base of a human-dominated hierarchy" (Hall 2011, 6). Rejecting what earlier plant studies scholars (Wandersee and Schussler 1999) rightly term 'plant-blindness' in Western culture, Gagliano condemns our current state of 'vegetal disregard' for "plants, whose fundamental role is to ensure continuity of life on Earth" (2013, 149). When we "contemplate and confront the vegetal in the human," Miller believes, we will advance the posthumanist project of overturning hierarchies, "strike at the root of humanity's instrumentalist domination of plants... [and] recognize kinship with plants [which] will inevitably alter how we think about our use of them" (462).

As with animal studies articulations of posthumanities, plant studies perspectives can be understood through their genealogies. The science studies perspectives in plant studies trace their field from 'electrifying discovery' in 1983, through 'decisive debunking' in 1984, and on to 'resurrection' by 1990 (McGowan 2013). Companioned by more popular science texts such as Michael Pollan's *The Botany of Desire* (2001) and Daniel Chamovitz' *What A Plant Knows: A Field Guide to the Senses* (2012), this science studies branch has gained academic attention through professional organizations such as the International Society of Plant Signaling and Behavior, whose annual conference in summer 2013 drew scholars from over 40 countries, and was discussed in the popular debut of plant studies (Pollan 2013). Prominent scholars such as Susan Murch (Canada Research Chair in Natural Products Chemistry), Catriona Sandilands (Canada Research Chair in Sustainability and Culture), science writer Elaine Dewar, and Basque philosopher Michael Marder each have plant studies research in press, with Dewar's *Smarts* expected in 2015, and Marder's new Rodopi Press series on "Critical Plant Studies: Philosophy, Literature, Culture" already producing its first volume on *Plants and Literature* (Laist 2013). Philosopher Michael Marder's *Plant-Thinking: A Philosophy of Vegetal Life* (2013) brings new materialism's concepts of agency and the posthumanities' decentering of the human to develop critical plant studies' redefinition of species as existing on a continuum that is more alike than different, with shared ancestry.

In articles introducing the new field of plant studies, both the purposes for plant studies research and the dominant standpoints seem overwhelmingly humanist and masculinist, instrumentalizing plants for new technologies, or theorizing about plants in ways that benefit humans. For example, Pollan (2013) cites a disproportionate number of male scientists (women number only seven out of 27 researchers cited), and gives 'poet-philosopher' Stefano Mancuso of the University of Florence's International Laboratory of Plant Neurobiology greatest prominence in philosophical discussions about how to interpret the science of plant studies. According to Mancuso, the reason to study plant behavior is because "we stand to learn valuable things and develop new technologies," perhaps "design better computers, or robots, or networks," harness plants for computational tasks or send them to other planets for exploration. Masculinist perspectives in plant science lead to questions that are limited and ultimately humanist: for example, observing 'interplant communication' through the release of volatile chemicals, ecologists Richard Karban and Martin Heil wonder "why should one plant waste energy clueing in its competitors about a danger?" and conclude that "plant communication is a misnomer" and should be called "plant eavesdropping" or even a "soliloquy" (McGowan 2013).

From a critical ecofeminist perspective, the plant studies genealogy as currently presented is a Euro-patrilineage that parallels the developments and omissions of animal studies (Gaard 2012). Presented as if the field emerged only recently, it erases not only the methodology and findings of Nobel prize-winner Barbara McClintock (Keller 1983) and marine biologist Rachel Carson (1951, 1962), along with two centuries of work on human-plant relations explored by women gardeners, scientific illustrators, animal writers, and ecological artists (Norwood 1993; Norwood & Monk 1987; Anderson 1991; Anderson & Edwards 2002; Gates 1998). Even more significantly, plant studies genealogies largely omit indigenous non-Western perspectives—contemporary ecoactivists and writers such as Winona LaDuke, Tom Goldtooth, Gloria Anzaldúa, Chico Mendez, Ken Saro-Wiwa, Wangari Maathai, and many others—whose cultures never made the Aristotelian divisions of humans from the rest of life, and thus whose writing about plants, animals, and ecology has never needed material feminism's recuperative concepts of 'transcorporeality' (Alaimo 2010) or 'naturecultures' (Haraway 2003) to illuminate *All Our Relations* (LaDuke 1999). Those plant studies scholars who do attend to indigenous perspectives currently work in the humanities wing of plant studies—anthropology (Kohn 2013; Viveiros de Castro 2004), philosophy (Hall 2011; Marder 2013), gender and cultural studies (Plumwood 2000, 2003, 2012; Sandilands 2014).

As usual in environmental studies, the findings of the environmental sciences wing are crucial yet insufficient, and may in fact be operating on distorted premises, purposes, or hypotheses; they lack the contextual, philosophical, and political reflectiveness of the environmental humanities. Bringing forward feminist animal studies in dialogue with Val Plumwood's critical ecofeminist work on indigeneity and vegetarianism, Hall's philosophical botany, and the queer/posthumanist/feminist approaches of Hall, Gagliano, and Sandilands is needed in cultivating a critical ecofeminist perspective on human, animal, plant, and ecological relations. Rethinking these relations augments our understanding of 'species' and 'meat.'

Defining 'Species' and 'Meat'

In animal studies, the key terms 'species' and 'meat' reference a wealth of field-defining and movement-building conceptual groundwork. Peter Singer's (1975) key concept of *speciesism* defined as "a prejudice or attitude of bias toward the interests of members of one's own species and against those of members of other species" (7) has been used to build analogies to other humanist structures of oppression—racism, sexism, classism, ethnocentrism, anthropocentrism—and to persuade progressive communities of feminists, civil libertarians, labor activists, antiracist allies and environmentalists alike to recognize and reject mutually-reinforcing forms of hierarchical and dominative thinking, or what ecofeminist philosopher Karen Warren first called the *logic of domination* (1990).[3] The year 1990 was equally significant for feminist animal studies: the leading journal of feminist scholarship, *Signs*, published Josephine Donovan's essay describing more than a century of vegetarian women's activism and theory connecting a feminist ethics of care with animal defense, and Carol Adams' *The Sexual Politics of Meat: A Feminist-Vegetarian Critical Theory* explored the ways gender dualisms linking men, meat-eating, and virility in opposition to women, vegetables, and passivity reinforce the subordination of women and non-human animals through processes of objectification, fragmentation, and consumption. Adams' concept of animals as the *absent referent* of meat eating, and female animals as the *absent referent* in not only meat eating but also dairy and egg production gave vegan feminist language to what it means to become 'a piece of meat.' Shortly afterwards, *Ecofeminism: Women, Animals, Nature* (Gaard 1993) became the first volume to link feminist animal studies with ecofeminism by placing species at the center of ecofeminist praxis, and launching a decade of debate among feminists and ecofeminists about the place of animals in an antiracist, non-essentialist and postcolonial ecofeminism (Gaard 2003, 2011).

In 2014, feminist animal studies still differs from the newer, yet mainstream branch of animal studies, and is more closely allied with critical animal studies. Whereas animal studies has tended to investigate human-animal relations from an academic perspective, both feminism and critical animal studies are movements for justice; many critical animal studies scholar-activists are also feminists. Unlike animal studies, at the heart of feminism is the centrality of praxis, the necessary linkage of intellectual, political, and activist work. Feminist methodology has challenged the male bias masquerading as objectivity in science, and worked to undermine the fit of science with dominant modes of exploitation and oppression that use science to benefit elite humans, often at the expense of disenfranchised humans and experimented-upon animals (Harding 1987, 1991; Keller & Longino 1996; Stanley 1990). In stark contrast, feminist methodology requires that feminist research puts the lives of the oppressed at the center of the research question, and undertakes studies, gathers data, and interrogates material contexts with the primary aim of improving the lives and the material conditions of the oppressed. When feminists attend to 'the question of the animal,' we do so from a standpoint that centers other

---

3 Warren did not include speciesism in that logic for another decade; see Warren 2000.

animal species, makes connections among diverse forms of oppression, and seeks to put an end to animal suffering—in other words, to benefit the subject of the research (Birke & Hubbard 1995).

On the surface, contemporary plant studies may share a commitment to plant well-being as well. Its scholarship challenges the definitions of 'species' and 'meat,' charging that animal studies scholarship (across the branches of animal studies) has only 'moved the line' of moral considerability, performing a humanist moral extensionism that includes other animal species but places plants outside the bounds of moral consideration—effectively, *treating plants like meat*. In animal studies, plants are "backgrounded," writes Houle (91–92), and Matthew Hall (2011) agrees, invoking Val Plumwood's (1993) theory of the Master Model construction of both the Master identity and a logic of domination (Warren 1990) that operates through Plumwood's five linking postulates of the Other's homogenization, hyperseparation, backgrounding, instrumentalism, and denied dependency. In short, animal studies may be liberating for animals, but oppressive for plants, thus perpetuating humanism. Not only are plants 'kin' to animals, but the presence of carnivorous plants, herbivorous animals, sea anemones, algae and fungi (Tsing 2012) which ambiguously display attributes of both plants and animals have all begun to blur the demarcation of 'species' dividing plants from animals, first asserted by Charles Darwin's theory of common descent in his *On the Origin of the Species* (1859). This radical continuity, transcorporeality and kinship across plant and animal species gives rise to several questions: first, as feminist methodology suggests, we can ask, do plants benefit from plant studies? And second, inspired by a queer feminist and posthumanist animal studies, we must ask, without creating a moral underclass, how do we make ethical food choices in light of the fact that all potential 'foods' are sentient beings? And finally, in what ways can a feminist methodology be used to study and improve conditions and interspecies relations among plants, humans, and animals, augmenting critical animal studies by responding to the findings and implications of plant studies scholarship?

In considering these questions, there are arguments critical animal studies and ecofeminist scholars would want to endorse, and others to avoid. For example, plant neurobiologist Stefano Mancuso believes that "because plants are sensitive and intelligent beings, we are obliged to treat them with some degree of respect," which means "protecting their habitats" and "avoiding practices such as genetic manipulation, growing plants in monocultures, and training them in bonsai" (Pollan 2013). This standpoint seems consonant with feminist animal studies arguments and feminist methodology. But this standpoint does not prevent humans from eating plants since *"plants evolved to be eaten,"* Mancuso asserts, given plants' lack of irreplaceable organs and modular structure. This argument is reminiscent of parallel justifications of human predation on other species (animals) and their *telos* (i.e., they are 'meant to be eaten'), whether based on biology (theirs or ours), culture, or need.

Reviewing the arguments for veganism, we find some of these apply also to plants.

1. *They do not want to be eaten.* Countless texts in animal studies confirm that animals do not want to be eaten: their behavior speaks their desire as they run away from hunters, fight against other predators, and struggle to free themselves from zoos, leg-hold traps, science experiments, and other forms of confinement (Hribal 2011). But it appears plants do not want to be eaten either: they give off chemical signals when attacked by insects, alerting other plants, and sometimes invoking predatory insects to feed on the attackers; plants also produce toxins altering a leaf's flavor or texture, making it less palatable and less digestible to herbivores (Angier 2013; Pollan 2013). In queer theory and in feminist theory alike, a primary consideration is *consent*: that is, if all parties do not consent to a specific behavior or relationship, this lack of consent signals potential exploitation, oppression, or otherwise ethically dubious relations at work.

2. *They feel pain.* Animals suffer and feel pain, and thus deserve not to suffer: Singer's utilitarian argument has powered animal rights for decades. Animals are also clearly subjects-of-a-life, as Tom Regan has argued: animals feel pain, emotions, and have a sense of selfhood that affirms their intrinsic value and gives them moral rights; they are not to be used as means to an end for others (Regan 1983). Plant studies confirms that plants give off volatile chemical signals when attacked, suggesting plants experience vegetal versions of fear and pain. Although plants lack a brain and nervous system, their documented behaviors suggest a level of plant intelligence and plant communications that is presently beyond our knowledge. Lack of information does not ethically justify behavior (though it may explain behavior) and does not provide foundation for causing fear and pain without consent.

3. *They have consciousness.* While the consciousness of most animal species involves a sense of selfhood that is simultaneously individual, familial, species-based and relational, plant studies research suggests that plants have consciousness too in that they make decisions based on environmental information, communicate (or 'signal') to other plants and insects, share nutrients, nurture their offspring.

Adapting to plant studies and the questions of human diet the four frequently-invoked bases for veganism—environmental, human health, world hunger, animal suffering—and placing these in conjunction with Deane Curtin's (1991) contextual moral vegetarianism[4] offers some clarity. As Curtin succinctly argues, "the reasons for moral vegetarianism may differ by locale, by gender, as well as by class"; as a "contextual moral vegetarian," Curtin "cannot refer to an absolute moral rule that prohibits meat-eating under all circumstances" (69). For "the point of a contextualist ethic is that one need not treat all interests equally as if one had no relationship to any of the parties" (70), so Curtin can envision extreme situations of starvation or danger where killing another animal (of any species, including human) would be ethical. Curtin also cites environments that do not support a vegan or vegetarian diet as contexts where food ethics will be based on other considerations of relationships, both animal and ecological; he discusses cultures that pay respect to animals unavoidably killed during human agricultural practices (Japanese Shintoism), and cultures that give human bodies back to other animals as food, recognizing our place in the food chain (Tibetans). Instead of using other cultures' food practices as an excuse for Western predation, Curtin emphasizes, "if there is any context... in which moral vegetarianism is completely compelling as an expression of an ecological ethic of care, it is for economically well-off persons in technologically advanced countries" (70). To support this view, he cites the impact of first-world agriculture and consumption practices on first-world and global environments, the role animal agriculture plays in exacerbating human hunger, the exploitation of female animals in the egg and dairy industries, and the fact that first-world consumers have a variety of food options that do not perpetuate suffering. It is a moral and political injunction to "eliminate needless suffering wherever possible," Curtin argues (70). His argument has been further supported by the United Nations' Food and Agricultural Organization's Report, *Livestock's Long Shadow* (2006), documenting the deleterious ecological effects of industrial animal agriculture, and its link to climate change;

---

4   In distinguishing uses of 'vegetarian' and 'vegan' in the 1990s, I quote Richard Twine's (2014, 206n4) excellent explanation: "During the 1990s... the term 'vegetarianism' was mostly used instead of 'veganism.' I would contend three reasons for this. First, I expect some North American writers used vegetarianism but meant veganism. Secondly, since the 1990s... vegetarianism has lost a lot of credibility as a consistent ethical position within the animal advocacy movement but at *that* time it was still deemed credible. Thirdly, and relatedly, during the first decade of the twenty-first century, notably in Western countries, there has been an ethical shift toward, and cultural normalization of, veganism as the preferred and more consistent practice of animal advocates. So much so, that an ecofeminist arguing today for ovo-lacto vegetarianism would suffer from a credibility problem."

these findings have been popularized and presented in documentaries such as "Meat the Truth" (2008), "Beef Finland" (2012) and "Cowspiracy" (2014). Ample evidence documents the fact that eating such animals involves eating the planet, since the production of animals for food requires vast amounts of water, plants, soil, and other animals (Pimentel & Pimentel 2003); it requires the exploitation of low-waged workers in horrific working conditions (Schlosser 2001) and has devastating impact on human health (Robbins 1987; Campbell and Campbell 2006). Ethical eating is not merely a question of eater and eaten, but a question of eater-eaten-environment—and environments are simultaneously ecological, sociocultural, and economic.

Curtin kindly affirms that there may be no moral *destination*, but rather a moral *direction* we can move in making decisions around what counts as food; I call this insight 'kind' because it resists the judgmental attitude omnivores ascribe to vegans, and makes visible the fact that all food production involves some death. As Lori Gruen (2014) elaborates,

> We can't live and avoid killing; this is something I think has been underexplored in vegan literature.... We harm others (humans and nonhumans) in all aspects of food production. Many are displaced when land is converted for agricultural purposes, including highly endangered animals like orangutans who are coming close to extinction as a result of the destructive practices used to produce palm oil, a ubiquitous ingredient found in a large number of prepared 'vegan' food products. (132–133)[5]

Contextual moral veganism moves first-world consumers in a moral direction but does not eliminate the deaths and consumption of some sentient others, both animal and plant, who are more kin to humans than Westerners have recognized. An important component of critical ecofeminism, contextual moral veganism is capable of acknowledging the sentience of plants and other ecological beings, and in diverse contexts, placing humans in the food chain as both eater and eaten, pointing to context-specific moral directions that strive to produce the least suffering and greatest care for all involved: humans (industrial, rural, agricultural, indigenous), animals, plants, ecological entities.

Though ecofeminism has conceived of humans as always embedded within specific and diverse environments, developing theory in consideration of

---

5 Unsustainable palm oil requires deforestation and the deaths of orangutans alike, and a strong resistance is organized through The Orangutan Project (http://www.orangutan.org.au/index.htm). Thanks to Kate Rigby for alerting me to this important intersectional activism.

both social justice and environmental concerns, contemporary plant studies and new materialism's concepts of agency and transcorporeality raise more specific questions about the place of plants in environmental and dietary ethics alike. A paradigm-fracturing shift is needed here, one that acknowledges human inter-identity, inextricable from and supported by a web of relations with sentient, intelligent kin across species. Val Plumwood argued for such a shift in her theory of 'ecological animalism' (2012) or 'animist materialism' (2009) or what she variously calls a "critical feminist-socialist ecology" (2000, 285) and "critical ecological feminism" (2000, 289). I choose the term 'critical ecofeminism' both to acknowledge and to advance upon the foundational work bridging animal and plant studies, feminism and ecology, first world and indigenous perspectives that is the prescient hallmark of Plumwood's thinking. It is unfortunate her work rejecting 'ontological veganism' (Plumwood 2000, 2003) misunderstands, misrepresents, and omits significant arguments for contextual moral veganism initially advanced by Curtin (1991), and mounts a 'straw woman' argument against the work of vegan ecofeminists Carol Adams (1990, 2010) and Marti Kheel (2008) as ethnocentric 'universalism' that places humans outside of nature and separate from the food chain (Twine 2014). A strong antiracist ally committed to challenging white privilege, Plumwood misread the vegan feminism of Adams and Kheel as devaluing the ethics and worldview of indigenous Australians, whose diet (like all indigenous groups) has developed in relationship to their immediate environments, and includes a range of plants and animals, all of whom they regard as kin. In her theory of 'ecological animalism,' Plumwood (2003) advocated a "context-sensitive semi-vegetarian position" and opposed factory farming but not a subsistence, need-based killing of other animals; the goal of her theory was "situating human life in ecological terms, and situating non-human life in ethical terms." Because her unfounded attack on Adams and Kheel has been soundly critiqued by Richard Twine (2014), here, I will salvage those aspects of Plumwood's theory that are helpful in developing a critical ecofeminism that is responsive to agency and transcorporeality across plant and animal species.

As if anticipating critiques of the humanism in Ursula Heise's (2008) 'eco-cosmopolitanism,' Plumwood (2003) argues against the "'biosphere person' [whether vegan or omnivore] who draws on the whole planet for nutritional needs defined in the context of consumer choices in the global market" and whose lifestyle is "destructive and ecologically unaccountable." Instead, Plumwood develops a 'critical bioregionalism' (2008) that makes visible "the shadow places of the consumer self," those "places that take our pollution and dangerous waste, exhaust their fertility or destroy their indigenous or nonhuman populations in producing our food." In delightful anticipation of queering ecofeminism and plant studies, Plumwood proposes "envisioning a less monogamous ideal and a more multiple relationship to place" which she phrases as an "accountability requirement" involving "an injunction to cherish and care for your places, but without in the process destroying or degrading any other places, where 'other places' includes other human places, but also other species' places." Urging that we "try to see creativity and agency in the other-than-human world around us" (2003), Plumwood simultaneously maintains that "all embodied beings are food and more than food" and yet "no being, human or nonhuman, should be ontologised reductively as meat" (2009). In these statements, Plumwood develops a contextualized ethic of human inter-identity as embedded with ecological others, accepting our place in the food chain as both eater and eaten.[6] Considering ourselves as potential prey for other animals, but in the midst of our lives and after our death, is wholly consonant with Plumwood's critical ecofeminism (1995, 2012).

Westerners are troubled by indigenous views of non-human animals, plants, and ecological beings such as rocks, water, and soil as not only sentient, but kin to humans; seen as animism, these views suggest a perspective belonging to the disciplines of anthropology and comparative religions rather than environmental science. Yet, new developments in plant studies and elemental ecocriticism (Cohen 2010) recognize a similar animacy and agency in ecological others. In eco-anthropology, scholars such as Eduardo Viveiros de Castro (2004) and Eduardo Kohn (2013) describe the ecological worldview and linked self-identity of indigenous Amazonian cultures, which emphasize nondifferentiation between humans and animals, intercommunicability, and a state of being wherein self and other interpenetrate. As Viveiros de Castro explains, "cultivated plants may be conceived as blood relatives of the women who tend them, game animals may be approached by hunters as affines, shamans may relate to animal and plant spirits as associates or enemies" (466). In such worldviews, humans and animals are interchangeable, becoming one another as a result of death and consumption. For Westerners, such eating of one's kin seems like cannibalism, and indeed some vegan and vegetarian activists

---

6 The new materialisms have uncovered the various ways that our own bodies are colonized by microbial life, and how illnesses soon to be exacerbated by climate change are a manifestation of other microbial species feeding on humans. The hegemony of humanism prevents non-indigenous humans from conceiving of ourselves as prey, as food, for the duration of our lives, and only after death do some cultures (i.e. Tibetan) offer human bodies as food for other animals. The 'eater and eaten' phrase in the text is thus still unequal: over the course of a lifetime, humans do far more eating than being eaten.

consider eating other animal species a form of cannibalism.[7] But if plants are also sentient and kin, then all eating becomes the eating of relatives, a significant ethical conundrum for Westerners.

Horror movies "reflect a deep-seated dread of becoming food for other forms of life," Plumwood argues, but "as eaters of others who can never ourselves be eaten in turn by them or even conceive ourselves in edible terms, we take, but do not give" (2003). Rejecting this anti-ecological human supremacy, Plumwood argues that "in a good human life we must gain our food in such a way as to acknowledge our kinship with those whom we make our food, which does not forget the more than food that every one of us is, and which positions us reciprocally as food for others" (2003). But this reciprocity to the ecological network or system does not reciprocate to the individual being consumed, as Ralph Acampora (2014) argues. Taking Analía Villagra's work on "Cannibalism, Consumption, and Kinship in Animal Studies" (2011) seriously, Acampora explores her proposition that indigenous worldviews "allow for the consumption of fellow animals not in the absence of or in spite of bonds of kinship, but rather because of them" and thus Villagra argues for "becoming cannibal" (50, 52). But as Acampora observes, Villagra's indigenous cannibalism applies to the extended kinship of other species, and does not commend anthropophagy (cannibalism of humans by humans). Whereas Villagra's use of indigenous worldviews leads her to a moral relativism in her concluding discussion of "my delicious pet," Acampora emphasizes the fact that the indigenous worldviews Villagra discusses have clear understanding of kinship distances. It appears there is both a moral direction and moral destination in these contextual and relational practices of eating, living together, and sharing souls. We can honor all our relations and still move in a moral direction that reduces suffering across species and bioregions, though our lives will never reach a moral destination of universal non-harming.

Beyond Meat, Beyond Species

What shall we eat? Eating plants ensures humans *consume less*—less plants, and less animals—and ensures we cause less suffering for plants, animals, ecosystems, and other humans as well. It ensures we free up more land for all of

---

7   These are private conversations; to my knowledge, no one has yet theorized this connection besides Ralph Acampora (2014). I first noticed this association when indexing my first book (Gaard 1993), and discovering the term 'cannibalism' cropped up enough to be indexed; at the time, the prospect seemed too far-fetched to mention.

life—for plants, for animals, for humans to eat and thrive. And, it monkey-wrenches climate change. Bringing into dialogue the critical animal studies and plant studies branches of posthumanism through a lens that recuperates the feminist lineage of these branches, critical ecofeminism has much to offer.

It argues that undoing the grasp and hegemony of a carnist culture requires shifting from denial to attentive listening, from alienation to empathy, from capitalist production time to seasonal time, from a heteronormative universalism to a queer multiversalism. It involves a refiguration of selfhood from rationalist individualism to material transcorporeality, releasing the internalized capitalism of a self-worth based on ceaseless production, and replacing it with a selfhood attuned and intra-active with the cycles of seasonal growth and decay; it offers an acceptance of our death as part of these cycles, and locates our place in the food chain as both eaters and eaten, invoking a contextual moral veganism that values individual lives as well as ecosystem relations, and makes food choices that move in directions promoting sustainability and reducing suffering for eater-eaten-ecosystem.

Philosophers, eco-anthropologists, ecocritics and other scholars of Western culture are using the tools we have inherited as ways to theorize our understanding of other beings, of non-human animals, plants, minerals, and other planetary entities. But as Audre Lorde has written, "the master's tools will never dismantle the master's house" (1984). What's needed is a conceptual shift, a 're-think' as Plumwood proposed (2009), re-situating our perspectives so that they companion the standpoints of non-Western cultures, other animals and plants. This shift involves the practice of attentive listening advocated by feminist animal studies scholars (Donovan 1990, 1998) and plant studies scholars as well: it involves learning a different language of embodiment, behavior, scent, and intra-activity. As strategies for seeing "creativity and agency in the other-than-human world around us" (Plumwood 2009), Westerners can stand with the struggles and lives of animals, plants, and indigenous peoples; we can utilize tools from the Buddhist practice of mindfulness and the non-theistic principles of the dharma, originating in India but freely offered across cultures.

From Plato and Descartes, Westerners have learned to "treat consciousness rather than embodiment as the basis of human identity" (Plumwood 2004, 46). But these elements of being are inseparable, intra-active elements of agency, interconnected with other flows of planetary life. Just as animal studies loosened the grip of humanism for Westerners, now plant studies—along with vegetal and elemental ecocriticism—offer additional tools. For example, Michael Marder's *Plant-Thinking* (2013) proposes that we examine the world of plants from their own perspective, a practice of attentiveness that uncovers the meaning of plant life, made evident through the seasonal changes, growth, and cyclical character of being. To cultivate this attentive stance, Plumwood (2005) proposes a practice of decolonizing gardening, "a healthful pursuit that brings gardeners into contact and collaboration with nature and sensitizes them to the earth, the rhythms of the seasons, growth processes and the life and death cycles of living things." When Marder discusses the freedom of plants, he argues that plants' indifference, lack of selfhood, and lack of concern for their own self-preservation make them capable of the freedom of play; thus, attentiveness to plants leads not only to plant emancipation, but our own alongside, enabling us to recognize the rich diversity of perspectives possible when consciousness and thought are understood as a creative and inventive "thinking before thinking" (154). Plants' lack of selfhood echoes the Buddhist no-self (*anatta*), a concept linked to awareness of impermanence (*anicca*) and dependent origination (*paṭiccasamuppāda*), the understanding that all beings and events co-arise and pass away. Plants' alleged lack of concern for self-preservation (not entirely accurate if we consider the volatile scents released when plants perceive an attack) shows up in the Buddhist concept of non-clinging, since clinging (*tanha*) is the basis of suffering (*dukkha*).

From a Buddhist perspective, the alleged indifference of plants, like the imputed indifference of rocks and other minerals, might be better understood as *equanimity*, the ability to be present to life without placing conditions on how life shows up. In ecofeminist theory, Karen Warren's (1990) work demonstrates how the logic of domination has been used to rationalize human dominance over plants and rocks (128–129), a dominance she refutes through a rock-climbing narrative that offers a Westerner's practice of attentiveness to mineral life (cf. Cohen 2010; Gaard 2007), leading to a recognition that ethical practice depends not on the quality or attitude of the other, but rather on the quality of relationship and attention humans bring to regard the other:

> I closed my eyes and began to feel the rock with my hands—the cracks and crannies, the raised lichen and mosses, the almost imperceptible nubs that might provide a resting place for my fingers and toes when I began to climb. At that moment I was bathed in serenity.... I felt an overwhelming sense of gratitude for what [the rock] offered me—a chance to know myself and the rock differently, to appreciate unforeseen miracles like the tiny flowers growing in the even tinier cracks in the rock's surface, and to come to know a sense of being in relationship with the natural environment. It felt as if the rock and I were silent conversational partners in a longstanding friendship. (Warren 1990, 134)

Attending to animals, plants, minerals and other planetary life depends not only on the inclusion of others but on the quality of attention to the relations.

The emergence of plant studies illuminates the vegetal considerations already present in indigenous cultures, in Buddhism, and in critical ecofeminism, a Western perspective that brings together social and environmental justice, climate justice, and interspecies justice. Ecofeminism's contextual moral veganism offers a useful strategy for making decisions about ethical eating for humans, plants and animals; its contextual aspect is not a form of moral relativism, nor is it a universal rule. A critical ecofeminism encourages a shift in our thinking and in our being, from a humanist perspective of dominance to an awareness and participation in relations of mutuality and reciprocity that resituate humans in the cycles of planetary life.

References

Acampora, Ralph. "Caring Cannibals and Human(e) Farming: Testing Contextual Edibility for Speciesism." In *Ecofeminism: Feminist Intersections with Other Animals and the Earth*, edited by Carol J. Adams and Lori Gruen, 144–158. New York: Bloomsbury Publishing, 2014.

Adams, Carol J. *The Sexual Politics of Meat: A Feminist-Vegetarian Critical Theory*. New York: Continuum Press, 1990 & 2010.

Adamson, Joni, and Cate Sandilands. "Vegetal Ecocriticism: The Question of 'The Plant.'" Pre-Conference Seminar for ASLE 10th Biennial Conference, 2013. http://asle.ku.edu/Preconference/adamson-sandilands.php.

Alaimo, Stacy. *Bodily Natures: Science, Environment, and the Material Self*. Bloomington, IN: Indiana University Press, 2010.

Anderson, Lorraine. (ed.) *Sisters of the Earth: Women's Prose and Poetry About Nature*. New York: Random House/Vintage Books, 1991.

Anderson, Lorraine, and Thomas S. Edwards. (eds.) *At Home on This Earth: Two Centuries of U.S. Women's Nature Writing*. Hanover: University Press of New England, 2002.

Angier, Natalie. "How Plants Defend Themselves: Maybe it's Time to Reconsider What We Think about Plants—They Hear, Touch, See and Even 'Talk' in Order to Survive." *Mother Earth News*, April/May, 2013. 63–64, 66–67.

Armbruster, Karla. "'Buffalo Gals Won't You Come Out Tonight': A Call for Boundary-Crossing in Ecofeminist Literary Criticism." In *Ecofeminist Literary Criticism*, edited by Greta Gaard and Patrick Murphy, 97–122. Urbana, IL: Illinois University Press, 1998.

Bekoff, Marc, and Jane Goodall. *The Emotional Lives of Animals*. Novato, CA: New World Library, 2007.

Bekoff, Marc, and Jessica Pearce. *Wild Justice: The Moral Lives of Animals*. Chicago: University of Chicago Press, 2009.

Birke, Lynda, and Ruth Hubbard. (eds.) *Reinventing Biology*. Bloomington, IN: Indiana University Press, 1995.

Browne, Kath, and Catherine J. Nash. (eds.) *Queer Methods and Methodologies*. Surrey, England: Ashgate Publishing, 2010.

Campbell, T. Colin, and Thomas M. Campbell. *The China Study*. Dallas, TX: Benbella Books, 2006.

Carson, Rachel. *Silent Spring*. Boston: Houghton Mifflin Company, 1962.

Carson, Rachel. *The Sea Around Us*. Oxford University Press, 1951.

Chamovitz, Daniel. *What a Plant Knows: A Field Guide to the Senses*. New York: Scientific American / Farrar, Straus and Giroux, 2012.

Chang, Chia-ju. "Putting Back the Animals: Woman-Animal Meme in Contemporary Taiwanese Ecofeminist Imagination." In *Chinese Ecocinema in the Age of Environmental Challenge*, edited by Sheldon Lu, 255–270. China: Hong Kong University Press, 2009.

Cohen, Jeffrey Jerome. "Stories of Stone." *Postmedieval: A Journal of Medieval Cultural Studies* 1.1/2 (2010): 56–63.

Coole, Diana, and Samantha Frost. (eds.) *New Materialisms: Ontology, Agency, and Politics*. Chapel Hill, NC: Duke University Press, 2010.

Curtin, Deane. "Toward an Ecological Ethic of Care." *Hypatia* 6.1 (1991): 60–74.

Darwin, Charles. *On the Origin of the Species*. London: John Murray, 1859.

Derrida, Jacques. "The Animal That Therefore I Am (More to Follow)." *Critical Inquiry*, 28.2 (2002): 369–418.

Dewar, Elaine. "Shh...The Plants are Thinking." *Macleans* (Canada), September 16, 2013. http://www.macleans.ca/society/life/shh-the-plants-are-thinking/

Donovan, Josephine. "Tolstoy's Animals." *Society and Animals* 17 (2009): 38–52.

Donovan, Josephine. "Ecofeminist Literary Criticism: Reading the Orange." In *Ecofeminism: Women, Animals, Nature*, edited by Greta Gaard. Philadelphia: Temple University Press, 74–96, 1998.

Donovan, Josephine. "The Pattern of Birds and Beasts: Willa Cather and Women's Art." In *Writing the Woman Artist*, edited by Suzanne Jones, 81–95. Philadelphia: University of Pennsylvania Press, 1991.

Donovan, Josephine. "Animal Rights and Feminist Theory." *Signs* 15.2 (1990): 350–375.

Estok, Simon. "Theorizing in a Space of Ambivalent Openness: Ecocriticism and Ecophobia." *Interdisciplinary Studies in Literature and Environment* 16.2 (2009): 203–226.

Food and Agricultural Organization of the United Nations. *Livestock's Long Shadow*. Rome, Italy: FAO, 2006.

Gaard, Greta. "Literary Milk: Breastfeeding Across Race, Class, and Species in Contemporary U.S. Fiction," *Journal of Ecocriticism* 5.1 (2013): 1–18.

Gaard, Greta. "Speaking of Animal Bodies." *Hypatia* 27.3 (2012): 29–35.

Gaard, Greta. "'Ecofeminism' Revisited: Rejecting Essentialism and Re-Placing Species in a Material Feminist Environmentalism." *Feminist Formations* 23.2 (2011): 26–53.

Gaard, Greta. *The Nature of Home: Taking Root in a Place*. Tucson, AZ: University of Arizona Press, 2007.

Gaard, Greta. "Vegetarian Ecofeminism: A Review Essay." *Frontiers* 23.3 (2003): 117–146.

Gaard, Greta. "Tools for a Cross-Cultural Feminist Ethics: Ethical Contexts and Contents in the Makah Whale Hunt." *Hypatia* 16.1 (2001): 1–26.

Gaard, Greta. "Strategies for a Cross-Cultural Ecofeminist Ethics: Interrogating Tradition, Preserving Nature in Linda Hogan's *Power* and Alice Walker's *Possessing the Secret of Joy*." *The Bucknell Review* 44.1 (2000): 82–101.

Gaard, Greta. (ed.) *Ecofeminism: Women, Animals, Nature*. Philadelphia: Temple University Press, 1993.

Gagliano, Monica. "Seeing Green: The Re-discovery of Plants and Nature's Wisdom." *Societies* 3 (2013): 147–157.

Gates, Barbara T. *Kindred Nature: Victorian and Edwardian Women Embrace the Living World*. Chicago: University of Chicago Press, 1998.

Gleig, Ann. "Queering Buddhism or Buddhist De-Queering? Reflecting on Differences Amongst Western LGBTQI Buddhists and the Limits of Liberal Convert Buddhism." *Theology & Sexuality* 18.3 (2012): 198–214.

Gruen, Lori. "Facing Death and Practicing Grief." In *Ecofeminism: Feminist Intersections with Other Animals and the Earth*, edited by Carol J. Adams and Lori Gruen, 127–141. New York: Bloomsbury Press, 2014.

Hall, Matthew. *Plants as Persons: A Philosophical Botany*. Albany, NY: SUNY Press, 2011.

Haraway, Donna. *The Companion Species Manifesto*. Chicago: Prickly Paradigm Press, 2003.

Harding, Sandra. (ed.) *Whose Science? Whose Knowledge?* Ithaca, NY: Cornell University Press, 1991.

Harding, Sandra. *Feminism and Methodology*. Bloomington, IN: Indiana University Press, 1987.

Heise, Ursula. *Sense of Place, Sense of Planet*. Oxford University Press, 2008.

Houle, Karen L. F. "Animal, Vegetable, Mineral: Ethics as Extension or Becoming? The Case of Becoming-Plant." *Journal for Critical Animal Studies* 9.1/2 (2011): 89–116.

Hribal, Jason. *Fear of the Animal Planet: The Hidden History of Animal Resistance*. AK Press, 2011.

Jagose, Annamarie. *Queer Theory: An Introduction*. New York: NYU Press, 1996.

Joy, Melanie. *Why We Love Dogs, Eat Pigs, and Wear Cows: An Introduction to Carnism.* San Francisco, CA: Conari Press, 2010.

Keller, Evelyn Fox. *A Feeling for the Organism: The Life and Work of Barbara McClintock.* New York: W. H. Freeman & Company, 1983.

Keller, Evelyn Fox, and Helen Longino. (eds.) *Feminism and Science.* Oxford University Press, 1996.

Kheel, Marti. *Nature Ethics: An Ecofeminist Perspective.* Lanham, MA: Rowman & Littlefield, 2008.

Kohn, Eduardo. *How Forests Think: Toward an Anthropology Beyond the Human.* Los Angeles: University of California Press, 2013.

LaDuke, Winona. *All Our Relations.* Boston: South End Press, 1999.

Laist, Randy. *Plants and Literature: Essays in Critical Plant Studies.* Amsterdam: Rodopi, 2013.

Lorde, Audre. *Sister Outsider.* New York: The Crossing Press Feminist Series, 1984.

Marder, Michael. *Plant-Thinking: A Philosophy of Vegetal Life.* New York: Columbia University Press, 2013.

McGowan, Kat. "The Secret Language of Plants." *Quanta Magazine,* December 16, 2013. http://www.simonsfoundation.org/quanta/20131216-the-secret-language-of-plants/.

Miller, T. S. "Lives of the Monster Plants: The Revenge of the Vegetable in the Age of Animal Studies." *Journal of the Fantastic in the Arts* 23.3 (2012): 460–479.

Mortimer-Sandilands, Catriona, and Bruce Erickson. (eds.) *Queer Ecologies: Sex, Nature, Politics, Desire.* Bloomington, IN: Indiana University Press, 2010.

Norwood, Vera. *Made From This Earth: American Women and Nature.* Chapel Hill: University of North Carolina Press, 1993.

Norwood, Vera, and Janice Monk. (eds.) *The Desert is No Lady: Southwestern Landscapes in Women's Writing and Art.* New Haven: Yale University Press, 1987.

Pimentel, David, and Marcia Pimentel. "Sustainability of Meat-Based and Plant-Based Diets and the Environment." *American Journal of Clinical Nutrition* 78.3 (2003): 660S–663S.

Plumwood, Val. *The Eye of the Crocodile.* Edited by Lorraine Shannon. Canberra: Australian National University E-Press, 2012.

Plumwood, Val. "Nature in the Active Voice." *Ecological Humanities* 46. May, 2009. http://www.australianhumanitiesreview.org/archive/Issue-May-2009/plumwood.html

Plumwood, Val. "Shadow Places and the Politics of Dwelling." *Australian Humanities Review* 44, 2008. http://www.australianhumanitiesreview.org/archive/Issue-March-2008/plumwood.html

Plumwood, Val. "Gender, Eco-Feminism and the Environment." In *Controversies in Environmental Sociology,* edited by Robert White, 43–60. Cambridge University Press, 2004.

Plumwood, Val. "Animals and Ecology: Towards a Better Integration." Working/Technical Paper. Australian National University Digital Collection, 2003. http://hdl.handle.net/1885/41767

Plumwood, Val. "Integrating Ethical Frameworks for Animals, Humans, and Nature: A Critical Feminist Eco-Socialist Analysis." *Ethics and the Environment* 5.2 (2000): 285–322.

Plumwood, Val. "Human Vulnerability and the Experience of Being Prey." *Quadrant* 39.3 (1995): 29–34.

Plumwood, Val. *Feminism and the Mastery of Nature*. New York: Routledge, 1993.

Pollan, Michael. "The Intelligent Plant: Scientists Debate a New Way of Understanding Flora." *The New Yorker*, December 23, 2013. http://www.newyorker.com/reporting/2013/12/23/131223fa_fact_pollan?currentPage=all.

Pollan, Michael. *The Omnivore's Dilemma*. New York: Penguin Books, 2006.

Pollan, Michael. *The Botany of Desire*. New York: Random House, 2001.

Regan, Tom. *The Case for Animal Rights*. Berkeley, CA: University of California Press, 1983.

Robbins, John. *Diet For a New America*. Walpole, NH: Stillpoint Publishing Co, 1987.

Ryan, John Charles. "Passive Flora? Reconsidering Nature's Agency through Human-Plant Studies." *Societies* 2.3 (2012): 101–121.

Sandilands, Catriona. "Botanically Queer." Presentation for the Institute for Research on Women, Gender, and Sexuality at Columbia University, March 27, 2014. http://vimeo.com/90535517.

Schlosser, Eric. *Fast Food Nation*. New York: Houghton Mifflin Co, 2001.

Singer, Peter. *Animal Liberation*. New York: Avon Books, 1975.

Stanley, Liz. *Feminist Praxis*. London: Routledge, 1990.

Tompkins, Peter, and Christopher Bird. *The Secret Life of Plants*. New York: Harper Perennial, 1973.

Tsing, Anna. "Unruly Edges: Mushrooms as Companion Species." *Environmental Humanities* 1 (2012): 141–154.

Twine, Richard. "Ecofeminism and Veganism: Revisiting the Question of Universalism." In *Ecofeminism: Feminist Intersections with Other Animals and the Earth*, edited by Carol J. Adams and Lori Gruen, 191–207. New York: Bloomsbury, 2014.

Villagra, Analía. "Cannibalism, Consumption, and Kinship in Animal Studies." In *Making Animal Meaning*, edited by Linda Kalof and Georgina Montgomery, 45–58. Michigan State University Press, 2011.

Viveiros de Castro, Eduardo. "Exchanging Perspectives: The Transformation of Objects into Subjects in Amerindian Ontologies." *Common Knowledge* 10.3 (2004): 463–484.

Wandersee, James H., and Elisabeth E. Schussler. "Preventing Plant Blindness." *American Biology Teacher* 61.2 (1999): 82–86.

Warren, Karen J. "The Power and the Promise of Ecological Feminism." *Environmental Ethics* 12.2 (1990): 125–146.

Warren, Karen J. *Ecofeminist Philosophy*. Boulder, CO: Rowman & Littlefield, 2000.

Wolfe, Cary. "Human, All Too Human: 'Animal Studies' and the Humanities." *PMLA: Publications of the Modern Language Association* 124.2 (2009): 564–575.

Index

abattoirs—see slaughterhouses
Ablach Farm (in *Under the Skin*)   152–153, 157
absent referent, the   22, 34–37, 42–43, 45, 54, 117, 172, 205, 208, 210, 272
Acampora, Ralph   279
Adams, Carol J.   2, 19, 31–53, 99, 102, 117, 151, 153, 172, 178, 205, 248, 261
advertising industry   23, 34, 35, 37, 52, 84, 90–105, 109, 113–114, 123, 160, 223, 225, 241
Africa, dairy and meat industry   4, 10
agency, in animals and plants   266–267, 270, 277–278, 280
Akabas, Sharon   92
aliens (in science fiction)   151–153, 198–202, 204, 206–211, 214–216
Almiron, Nuria   56
ambivalence, towards meat & pet food   232–233, 239–240
American Society for Literature and the Environment (ASLE)   264–265
animal abuse   22, 23, 73–78, 84, 156, 179, 180–181
animal activism   14, 22, 24, 42, 46–51, 86, 111, 165–166, 175, 178–180, 192, 247, 264–265, 272, 278
animal industrial complex   2, 27, 56, 63, 125, 246, 247, 250, 262, 264
animal nationalism   22, 82–83
animal rights   9, 45–51, 78, 83, 100, 111, 125, 152, 166–167, 174–175, 178–179, 192, 233, 264, 274
animal welfare   5, 48, 50, 54, 66, 68, 73–79, 81–84, 86, 113, 150, 155, 159, 165–167, 174, 177–178, 190, 194–195, 226, 230, 232–234, 236, 239
animalization of women   36–37
animals as subjects—in art   163–183
Animals Australia   73, 75
animism   267, 278
anthropocentrism   11, 26, 32, 33, 39, 41, 55, 67–69, 122, 178, 200, 212–214, 265–271
anthropomorphism   48, 52, 58, 59, 118, 119, 123, 192–195, 267

aquaculture   3, 11
Arla (Swedish dairy company)   114–117, 125
art and ethics   171–176
Artists and Animals Survey   163–183
artists—and animal rights   166–167, 174–175, 178–179
    and farm animals   24, 163–183
Asia, chicken meat production   14
Australia
    beef industry   82, 131, 133–146
    chicken consumption   5
    dairy industry   23, 82, 124, 131, 141–142
    fish consumption   10
    live export industry   22, 73–86
    slaughterhouses   73–79, 82, 124, 131, 133–142, 145
Baker, Steve   59, 160, 171–172, 180
'becoming plant'   269
beef (and beef industry)   4, 7–8, 17, 20, 34, 41, 44, 54–55, 64–67, 95–96, 116, 122, 131, 139, 146, 158, 169, 209, 233, 265, 276
beef contamination scare   22, 53–69
Berger, John   77, 172
Big Mac (burger)   91, 93
bobby calves, see calves
Borat (film)   98–99
'botanical queers'   269
bovine emotional labour   111, 118–121
Boyde, Melissa   23–24, 110, 129–148
Brazil, meat consumption   5, 14, 82
Breeze Harper, Amy   74
Britain—see United Kingdom
broiler chicks   10, 12–16, 150
'broiler hysteria'   14
BSE (Bovine Spongiform Encephalopathy)   63
Budeşti   93, 103
buddhism   280–282
buffalo, numbers slaughtered   3–4
Burger King   23, 37, 90–106
Burt, Jonathan   172–174
Butler, Judith   43

CAFO (Concentrated Animal Feeding Operations)   1, 2, 6, 7, 14, 23, 149–162
calves   7–10, 109, 111, 114–118, 122–123, 137, 142
camera obscura   131–132, 142
cannibalism   150, 278–279
Capaldi, Peter   198–199, 201, 217
Capitalism   2, 17, 20, 51, 250, 280
carbon impact   17, 243–244, 250, 262
carnism   2, 18–19, 21, 23, 25–27, 33–34, 36, 39, 41, 44, 51–52, 55, 67–69, 95, 111, 117, 150, 151, 156, 208, 222–224, 228–230, 232–240, 264–265, 280
carnivorous virility   34, 36, 41, 44, 52
carnophallogocentrism   22, 31–53
cattle—see cows
Chaudhuri, Una   119
chicken meals in *Doctor Who* (TV series)   208–209
chickens
    advocacy of   25, 184–197
    cognition of   13, 25, 173, 185–196
    consumption of   5, 12–16, 18
    farming of   2, 12–16, 194
    friendships of   14, 190–191
    global numbers slaughtered   3–4
    intensive farming of   12–16, 150
    male chicks, slaughter of   14
    personhood   25, 184–195
    relationships with each other   132–133, 142–144, 147
    removal from natural behaviors   12–16
    sentience of   12–13, 173, 185–196
    slaughter   16
    trivialization of   12
child-animal relationships (and Swedish dairy marketing)   121–124
China
    chicken meat industry   14
    commercial fishing   11
    free trade agreement   82
    meat industry   4–5
    milk/dairy   113
Chocowinity Chicken Sanctuary   15
Christmas dinners, in *Doctor Who*   202, 207–212
class, and meat consumption   4, 10, 18, 55, 64, 96, 99–100, 198, 207, 231, 247, 249, 257, 268, 271, 273, 275

climate change   20, 62, 66, 243, 250, 275, 278, 280
cognition, ranking of animals according to   25, 73, 185–188, 194, 239
colonialism   10, 23, 80, 91, 94–97
colonization   78, 80, 94–95, 101, 278, 281
companion animals   24–25, 54, 57, 60, 65, 164–165, 168, 185–186, 190, 192, 222–224, 227–228, 230–233, 238–240
contamination, of meat   54–69
cows (and cattle in general)
    as companions   131–133, 142–147
    as 'machines' (in dairy industry)   114–117, 123
    Australian live export of   22–23, 73–86
    consumption of   5
    dairy (see dairy industry)
    farming of   2, 7–10
    free-living herd in Australia   131–132, 141–144, 147
    natural lifespan of   10
    natural lives of   7
    numbers slaughtered   3–4, 18
    relationships with each other   132–133, 142–144, 147
    Swedish dairy industry   23, 109–125
    Swedish pasture releases   109–125
Crary, Jonathan   131, 142
Critical Animal Studies   21
Critical Plant Studies   26
Cudworth, Erika   25

dairy industry   2, 4, 7–10, 14, 16–17, 23, 26, 82, 109–125, 131, 141, 142, 169, 211, 243, 247, 249–250, 257, 272, 275
Davies, Russell T.   205–209, 216
Derrida, Jacques   22, 31–53
Dewar, Elaine   269–270
disgust   48, 57, 149, 228–229, 234–237
distancing (of consumer from meat production process)   152–156, 160
Doctor Who (TV series)   25, 198–221
'double-speak' (in meat industry)   150, 156
Douglas, Mary   22, 55, 58–61, 65, 68–69
Dracula   92–93, 100
Du Plessis, Rachel   140
Du Puis, E. Melanie   96
ducks, numbers slaughtered   3–4
dystopian fiction   161

INDEX   291

'Eating Well' (Derrida)   32, 33
Eccleston, Christopher   198–199, 201, 203, 209
ecofeminism   60, 100, 264–282
egg industry   12–16, 18, 84, 178, 227, 272, 275
embodied performance   246
empathy   1, 25, 67, 184, 188, 193–194, 215, 223, 229, 256, 280
environment, impacts of farming on   2, 5, 16–17, 20, 48, 51, 58, 64, 84, 86, 100, 106, 113, 121–122, 125, 166, 174–178, 239, 246, 271–278, 281
essentialism   235, 257, 268
ethnography   112, 224, 231, 250
European horsemeat scandal   22, 53–69
European Union   5, 10
exoticism   23, 93, 96–98, 100, 102, 120, 212
experimentation on animals   13, 85, 119, 185–186, 192, 202, 239, 272, 274

Faber, Michel   21, 23, 149–162
factory farming—see CAFOS
family relationships, & carnism & veganism   244, 246–248, 251–254, 256, 258
farm animals
    underrepresentation of, in art   163–183
    perceptions of as unnatural   169–170, 193
fast food   1, 16, 23, 91, 97, 99–100, 103–106
feminism   2, 22, 26, 33, 35, 40, 49, 50, 83, 99, 100, 175, 250, 264–282
fiction (and human-animal divide)   149–162
Fiddes, Nick   2, 55
fish
    asphyxiation   12
    consumption of   10–12
    Doctor Who (BBC TV series)   203, 210–215
    number eaten in one lifetime   18
    numbers slaughtered   3
    sentience of   11
    trawl nets   12
flexitarians   20
Fonterra   7
Foucault, Michel   111, 116, 123, 156, 176
France
    and veal production   10
    and mislabled meat products   54
Francione, Gary   46

Frankenstein's monster   265
free range farming   2, 84, 150, 166, 235–236
Fruno, Ashley   180

Gaard, Greta   26, 60, 265, 271–272, 279, 281
geese, numbers slaughtered   3–4
gender and meat consumption   18, 22, 33
genetic engineering   1, 2, 13, 122, 151, 156, 193–195, 273
Germany
    male chicks   14
    mislabelled meat products   54
Gigliotti, Carol   172, 180
global meat industry   1–7, 10–13, 17, 23, 26, 82–83, 86, 105–106, 176, 222, 227, 243, 275, 277
globalization   20–23, 54, 56, 82–83, 86, 94, 101, 113, 222, 226
goats, numbers slaughtered   3–4
Great Ape Project, the   184–186, 188–189
Greenland, in Burger King ad   23, 90, 92, 97, 102
guinea fowl, numbers slaughtered   3–4

hamburgers   90–106
Happy Meals   103
'Happy meat' discourse   23, 111, 125
Haraway, Donna   240, 271
Health
    horsemeat scandal   60, 63, 65–68
    risks associated with meat consumption   5, 18, 20, 63, 66–67, 226, 229, 276
Heidegger   32, 39, 40
heterosexual universalism   280
Homebush State Abattoir (Australia)   133–147
homonationalism   83
horses   6, 22, 54–69, 190, 213, 232, 238
horsemeat, European scandal   22, 53–69
Houle, Karen   266–269, 273
human-animal divide   26, 38, 59, 158, 175
human-animal relations   57, 163–164, 168, 175, 177–181, 272
Human-Animal Studies   2, 21, 161, 163, 175–177, 179, 272
human-bovine relationships   109–125
human-plant relationships   271
human subjectivity   32, 33
Human Trophic Level   19

humanitarianism (pseudo)—in Burger King ads 23, 91, 97, 103–106

Iams 223, 225–228
India
   commercial fishing 11
   dairy industry 10
   meat consumption 4–5, 10, 82
   slaughter 129
   widow sacrifice (*sati*) 80
indigenous peoples 35, 40–42, 51, 271, 276–280, 282
Indonesia
   commercial fishing 11
   consumption of chickens 14
   live export, and 22, 73–86
   slaughterhouses 22
Inghams' chicken meat supplier 75–76
'instantaneous maceration' (of male chicks) 14
intellectual disability 25, 185, 189–190
intensive farming 2, 4, 5–7, 11, 17–18, 24, 149–150, 155, 166, 223
Italy, and veal production 10

Japanese earthquake/tsunami (2011) 129–131
Japanese Shintoism 275
Joy, Melanie 19, 36, 67–69, 223, 228–229, 265

Kalvin (Skånemejerier bovine mascot) 116–117, 125
kangaroos, slaughter of 78
Kazakhstan 98–99
Kheel, Marti 277
Kreckler, Derek 131, 141–142, 144, 147
Kymlicka, Will 74, 175–176

language, and meat culture 36, 48, 67, 114, 123, 140, 150, 156–159, 272, 280
Laudan, Rachel 95
Lavin, Chad 226
layer hens 11–13
Levinas 32, 40, 43, 47
'Life of the Dairy Cow' 142
live exports (Australian meat industry) 73–86
   racialized nature of 79–86
Livestock's Long Shadow 17, 176, 275

Living Among Meat Eaters (Carol Adams) 46
logocentrism 32, 33, 39, 45–48
Lorde, Audre 280
love/eat distinction 223

McDonalds (fast food chain) 93–94, 103, 224
Mancuso, Stephano 270, 273–274
Maramureş 91–94, 104, 106
masculinity and meat consumption 20, 33–36, 41, 44, 45, 55, 94, 96, 99–102, 105, 246, 249
Meat Atlas 3–5, 17
meat culture
   and assumptions re civilization 56–60, 81–84, 134, 149
   and carnism 19
   definition of 19–20
   manifestations of 1–28
Meat Free Monday movement 20
meat industry 1–2, 6, 13–14, 16, 23, 82, 86, 116, 122, 137, 152, 161, 224
   and human workers 16
   racialized geopolitics 73–74
Meat—with care! (campaign) 137–138
media, influence of 16, 22–23, 34, 54, 56, 58–59, 61–65, 71, 73–77, 81, 90, 92, 97–100, 104, 113, 198–200, 205, 213, 229
media, horsemeat scandal 61–67
memorial(s)—and animal deaths 129–147
'Memory Salvage' project 129–131
Milk
   consumption 10
   in *Doctor Who* (TV series) 203–204, 207–211, 216
   see also dairy industry
Molloy, Clare 61, 55, 56, 59, 61, 116
Moral issues 25, 26, 35, 42, 46, 48, 54–59, 67, 69, 73, 75, 77, 83, 84, 95, 116, 120, 151, 180, 184, 193, 199–202, 207, 213, 215, 229, 246, 248, 250, 264–265, 267, 273–282
multi-species households 26

nationalism and meat 54, 55, 56, 62, 65, 66, 82–83
Native Americans, genocide of 35, 41
naturalization of meat 19, 91, 156–157, 223, 229, 234–235, 237, 239, 257

INDEX

nature/culture dichotomy   24, 164, 271
neoliberalism   22, 54, 91
Netherlands, and veal production   10
New Carnivore movement   36
new materialism   269–270, 277–278
New Zealand   7–9, 10, 16, 21
Nonhuman Rights Project, the   184, 194–195
nursery rhymes (and farm animals)   135–136

omnivores   18, 21, 25, 101, 230, 243, 250–252, 256–261, 264, 276–277
open farm events (in Sweden)   111–125
orientalism   81, 83–84
O'Sullivan, Siobhan   73, 76, 77

Pacelle, Wayne   49–50
Pachirat, Timothy   2, 16, 76, 150
Pakistan, dairy industry   10
Parry, Jovian   36
pasture releases   23, 111–125
paternalism (& Burger King)   96, 106
Pepperberg, Irene   186
performance (and Swedish pasture releases)   23, 109, 111, 116, 119, 125
personhood, and nonhuman animals   184–197
Peru, commercial fishing   11
pet food industry   25, 55, 222–240
    companies   222, 225–227, 232
    history of   222–227
    recall of pet food   223, 227–228
    testing on animals   225, 238
    vegetarian dog 'owners', and   223–224, 228–240
pets—see companion animals
phallocentrism   32, 33, 39
phallogocentrism   48–52
Phelan, Peggy   132–133, 140–142
photography—and animal lives   129–132, 139–147
pigeons   186–187, 229
pigs   2–7, 11, 19, 57, 69, 75, 169, 184–187, 202, 210, 213, 225, 228–229, 232–233
    cognition of   184–187, 190, 229
    consumption of   5, 18
    intensive farming   6
    natural lives   6
    numbers slaughtered   3
    trophic level of   19

Plant Studies   26, 264–287
plants—consciousness   266, 274
Plumwood, Val   265, 271, 273, 277–281
Pollan, Michael   107, 264, 266–267, 270, 273–274
Pornography of Meat, The   22, 37
posthumanism   32, 33, 245, 247, 264, 268–273, 280
practice theory   26, 243–249, 256, 260, 262
poultry—see chickens, ducks, geese, guinea fowl
poverty   81, 91, 98, 103–106
Proctor and Gamble   224, 226
Purity and Danger (Douglas)   22, 55, 58, 62

queer multiversalism   280
Queer Studies/politics   51, 83, 268–271, 273–274, 277, 280

rabbits   6, 18, 229, 232, 234
racism   22–23, 64–65, 73–86, 95, 97–98, 207, 264, 271–272, 277
religious influences   5, 11, 58, 60, 74, 120, 278
representation
    of cattle in meat/dairy industries   109–125, 129–147
    of farm animals in art   163–183
Rogers, Lesley   12, 173, 186–189
Romania, in Burger King ad   23, 90–94, 97–99, 102, 105

sanctuary (for cows)   24, 131–133, 141–142
Sandilands, Catriona   264, 269–271
Saturday Night Live   21, 92, 105
Save Animals from Exploitation (SAFE)   9, 151
School visits (to Sweden dairy farms)   109–114, 121–125
Science fiction   24, 149–162, 198–221
Scotland, in *Under the Skin*   23, 151–152
sexualization of animals   36–38
Sexual Politics of Meat, The   2, 22, 31–38, 40, 43–44, 46, 48, 51–52, 102, 265, 272
Sheep   3–8, 18, 29, 65, 75–78, 100, 109, 134, 136–137, 139, 153, 163, 173, 225, 228–229, 238
Shove, Elizabeth   244–248
Singer, Peter   25, 46, 184–186, 190, 194, 264, 271, 174

Skånemejerier 109, 112, 114, 116, 118
slaughter/slaughterhouses 5, 7–8, 14, 16, 22, 24, 59, 64, 68, 73–76, 79–82, 133–142, 145, 149–153, 156–160, 167, 205, 224–225
snapshots (and memory, 'trace') 129–133, 141
soy milk 4, 124
speciesism 54, 56, 68, 120, 200, 202, 212–215, 262, 265, 271–272
Spivak, Gayatri 79–80, 85
State Abattoir at Homebush 23, 133–145
steak (in *Doctor Who*) 204–207, 212, 214
stereotype(s) of
    effeminate rice eater 91, 94–101, 105
    minstrelsy 97–98, 105
    Transylvania 91–94
'Sunday roast' (as eating practice) 245–246
sustainability 26, 113, 243–244, 249, 262, 276, 280
Sweden
    dairy industry 23, 109–125
    dairy marketing 109–125
    export of milk/dairy 112
    mislabeled meat 54
    number of dairy cows per farm 112
    'The Way of the Milk' 114–117, 125

TARDIS 207, 210–211
taste, and meat consumption 18–19, 90, 161, 208–209
taste test, Burger King 23, 90–91, 93, 102–103, 111
Tennant, David 201, 207–209
Thailand, in Burger King ad 23, 90–91, 97, 102
The Animal That Therefore I Am (Derrida) 43
The Lost and Found Project 129–130, 146
The Old Cow Project 24, 129, 131–132, 144, 147
The Omnivore's Dilemma 101, 264
The Secret Life of Plants 264
The Ticking Clock Tenure (novel) 175
Timelord (*Doctor Who*) 198, 207, 214
trace (notion of) 129–132, 139–147
Transylvania, in Burger King ads 90–94, 97, 100–101
trophic levels 19
Tuan, Yi-Fu 224

Turkeys, numbers slaughtered 3, 18
turkey meals in *Doctor Who* (TV series) 208–212
Twine, Richard 2, 5, 17, 26, 63, 246, 250, 275, 277

Under the Skin (novel) 23, 149–162
United Kingdom
    and fish and chips 10, 210–211
    and vegans 222–242, 243–263
United Poultry Concerns 25
USA
    commercial fishing 11
    food imperialism 23, 98
    meanings of meat 20
    meat exports 82
    meat industry 3, 5, 7
    number of veg*ns 21
    pet food industry 222, 225, 227, 228
    slaughterhouses 16
    xenophobia 90–91, 96–97, 105

veal 4, 8–10, 139
vegan 'curious' 261
vegan ethics, in *Doctor Who* 25, 198, 200
vegan feminist ecocriticism 26, 272
vegan practice 26, 200, 244, 248–250, 256–258, 260–262
vegan research 249–251
vegan response to horsemeat scandal 57
veganism 2, 20–21, 25–26, 44–48, 56, 207, 213, 216, 244, 247, 249–252, 256–262, 274–280, 282
vegans, relationships 26, 244–249, 251, 254, 256–258, 260, 262
vegans, transition 26, 243–263
vegetal ecocriticism 26, 264–265
vegetarianism 2, 18–21, 25–26, 34, 44–47, 56–57, 66, 102, 151–152, 166, 170–171, 200, 204, 207–209, 215–217, 243, 248, 252–255, 260, 265, 271–272, 275–278
    in *Doctor Who* (BBC TV series) 200, 204, 207–209, 215–217
vertical integration systems 1, 56
violence, in carnist culture 19, 24, 33, 36–37, 42–43, 45, 74, 85, 114, 153, 156, 203, 205, 207, 210, 212–213, 257
virility, and meat eating 33–36, 41, 44, 52, 55

visibility (of abuse)   75–76
vivisection (in *Doctor Who*)   202, 213
vodsels (in *Under the Skin*)   151–159

Wadiwel, Dinesh   21, 22–23, 73–86, 176
Warren, Karen   272–273, 281
Western metaphysics   31–43, 265
whiteness   23, 35, 78–81, 84–85, 96–100, 198, 231, 251, 277
Whopper Virgins, the   23, 90–106

witnessing, of animal exploitation and death   23, 76, 84, 116–117, 129–147, 158, 216
Wolfe, Cary   268

xenophobia   22, 23, 54, 64, 90–91, 96–97, 105

Zammit-Lucia, Joe   171, 173, 179, 180
Zoocentrism   269
zooësis   23, 111, 119, 125

www.ingramcontent.com/pod-product-compliance
Lightning Source LLC
Chambersburg PA
CBHW071346290426
44108CB00014B/1451